WOWideas!

A collection of the

World's Greatest

or otherwise notable United States

Patents!

ISBN 0-7414-1731-6

Published by:

519 West Lancaster Avenue
Haverford, PA 19041-1413
Info@buybooksontheweb.com
www.buybooksontheweb.com
Toll-free (877) BUY BOOK
Local Phone (610) 520-2500
Fax (610) 519-0261

Printed in the United States of America

Printed on Recycled Paper

Published October 2003

Dedication

To family and friends, who encouraged our creative spirits and kept us on track with this many-weekends labor of love.

Contents

Preface

A love for ideas, especially WOWideas!™, and the process of invention led us to assemble this collection of United States patents.

It is our desire that you will explore, learn, and share with others these WOWideas!, wonderful creations of the unbounded human mind.

-Alexander Tourneu & Associates, LLC

The History of the United States Patent System

"The Congress shall have Power.... To promote the Progress of Science and useful Arts, by securing for limited Times to authors and Inventors the exclusive Right to their respective Writings and Discoveries..." -*U.S. Constitution, Article I, section 8*

In 1787 the framers of the Constitution recognized the importance of science and technology to the growth and development of the United States and granted Congress the power to pass laws relating to patents. The first patent law, enacted in 1790, established the Patent Board, composed of Thomas Jefferson, then Secretary of State, Henry Knox, Secretary of War, and Edmund Randolph, Attorney General.

Twelve years later, Congress established the Patent Office in the State Department, headed by the Superintendent of Patents. In 1836 Congress reorganized the Patent and Trademark Office and designated the Commissioner of Patents and Trademarks as the head official. The Patent and Trademark Office remained in the State Department until 1849, when it was transferred to the Department of the Interior, and in 1925 to the Commerce Department, where it is today.

In addition to establishing the office and department responsible for patents, Congress passed laws governing the subject matter for which a patent may be obtained and the conditions necessary for an invention or design to be patented. Among other things, a utility invention had to be useful, novel and unobvious to someone skilled in that particular field of study. No easy undertaking.

A patent grants the inventor "the right to exclude others from making, using, offering for sale or selling" the invention in the United States or importing the invention to the United States. Although a patent provides a monopoly for a period of years (current patents are for 20 years), it also requires the inventor to disclose the details of his or her invention to the public. These requirements ensure government patent officials, as well as inventors, mechanics, manufacturers, agents, and judges and juries in courts in which legal cases involving patents might be tried, will understand the invention.

As part of their disclosure, United States patent law requires applicants to submit drawings of their inventions and written descriptions called specifications. Among the wide ranging examples of patent drawings in this book are: Thomas Edison's electric lamp of 1880, the 1906 Wright brothers' flying machine, and the 1942 drawings of Chester Carlson's Electrophotography "copier". From 1836 until 1880, a model of the invention was also a mandatory part of the patent application.

This collection includes over 150 carefully selected patents, from the more than six million United States patents that have been issued since July 31, 1790... Thomas Jefferson would be proud!

What is Intellectual Property?

"A man's mind, once stretched by a new idea, can never go back to its original dimensions"
-Oliver Wendell Holmes, Jr.

Intellectual Property is imagination made real. It is the ownership of a dream, an idea, an improvement, an emotion we can touch, see, hear, and feel. It is an asset just like your home, your car, or bank account. Like other kinds of property, intellectual property needs to be protected from unauthorized use. If you are an intellectual property owner, you need to protect those rights. If you are a user, you should respect them. There are four common ways to protect different types of intellectual property:

PATENTS can provide rights up to 20 years for inventions in three broad categories:

Utility patents protect useful processes, machines, and articles of manufacture, and compositions of matter. Some examples: fiber optics, computer hardware, and medications.

Design patents guard the unauthorized use of new, original, and ornamental designs for articles of manufacture: The look of an athletic shoe, a bicycle helmet, the Star Wars® characters are all protected by design patents. Design patents last only 14 years.

Plant patents are the way we protect invented or discovered, asexually reproduced plant varieties: Hybrid tea roses, insect resistant corn and super-big tomatoes are all examples of plant patents.

TRADEMARKS protect words, names, symbols, sounds, or colors that distinguish goods and services. Trademarks, unlike patents, can be renewed forever as long as they are being used in commerce. The roar of the MGM Lion®, and the pink of the Owens-Corning Pink Panther® are familiar trademarks.

COPYRIGHTS protect authorship, such as writings, music, and works of art that have been tangibly expressed. The Library of Congress registers copyrights, which last the life of the author plus a considerable number of years thereafter. Gone With The Wind© (the book and the film), Beatles recordings, and video games are all works that are copyrighted.

TRADE SECRETS is information that companies keep secret to give them an advantage over their competitors. The formula for Coca-Cola® is the most famous trade secret.

Index

Index

INVENTOR	INVENTION	PATENT	DATE
Edison, Thomas Alva	Vote Recorder	90,646	1869
Edison, Thomas Alva	Phonograph	200,521	1878
Edison, Thomas Alva	Light Bulb	223,898	1880
Edison, Thomas Alva	Motion Picture Projector	493,426	1893
Einstein, Albert; et al	Refrigeration	1,781,541	1930
Eisenstadt, Marvin	Sweet N' Low®	3,625,711	1971
Englebart, Douglas	Computer Mouse	3,541,541	1970
Faget, Maxime; et al	Manned Space Capsule	3,093,346	1963
Farnsworth, Philo	Black and White Television System	1,773,980	1930
Fender, Clarence	Electric Guitar	2,741,146	1956
Fermi, Enrico; et al	Neutronic Reactor	2,708,656	1955
Fleming, Chet	Brain Life-Support	4,666,425	1987
Ford, Henry	Engine Carburetor	610,040	1898
Forrester. Jay	Computer Disk Storage	2,736,880	1956
Fraze, Ermal	Pop Top Soda Can	D195604	1963
Freeman, George	Wood Screw	29,963	1860
Fuller, Buckminster	Geodesic Dome	2,682,235	1964
Gatling, Richard	Machine Gun	36,836	1862
Gillespie, George	Congressional Medal of Honor	D37236	1904
Gillette, King	Safety-Razor	775,134	1904
Ginsburg, Charles; et al	Video Cassette Recorder (VCR)	2,956,114	1960
Glidden, Joseph	Barbed Wire	157,124	1874
Goddard, Robert	Rocket Engine	1,102,653	1914
Goldman, Sylvan	Shopping Cart	2,196,914	1940
Goodyear, Charles	Vulcanized Rubber	3,633	1844
Graham, John Jr	Seattle Space Needle®	3,125,189	1964
Greatbatch, Wilson	Cardiac Pacemaker®	3,057,356	1962
Greenwood, Chester	Ear Muffs	188,292	1877
Hall, Charles	Aluminum Process	400,665	1889
Head, Howard	Tennis Racket	3,999,756	1976
Headrick, Edward	Frisbee®	3,359,678	1967
Heinz, John	Vegetable Sorter	212,000	1879
Hewlett, William	Frequency Generator	2,268,872	1942
Hodges, Robert	Automobile Airbag	2,755,125	1956
Hoffmann, Felix	Aspirin	644,077	1900
Hollerith, Herman	Census Calculator	395,782	1889
Hollingshead, Richard	Drive-In Movie Theater	1,909,537	1933
Hooker, William	Mouse Trap.	528,671	1894
Houdini, Harry	Divers Suit	1,370,316	1921

Index

INVENTOR	INVENTION	PATENT	DATE
Hughes, Howard Sr	Drill Bit	930,758	1909
Hunt, Walter	Safety Pin	6,281	1849
Hyatt, John	Billiard Ball	50,359	1865
Ives, Frederick	Half-Tone Process	495,341	1893
Jackson, Peter; et al	Software	5,552,982	1996
Jacob, Mary Phelps	Brassiere	1,115,674	1914
Jacoby, Sidney	Smoke Detector	3,938,115	1976
James, Richard	Slinky®	2,415,012	1946
Jarvik, Robert	Artificial Heart	4,173,796	1979
Judson, Whitcomb	Zipper	504,038	1893
Karwowski, Joseph	Body Preservation	748,284	1903
Kilby, Jack; et al	Handheld Calculator	3,819,921	1974
Knabusch, Ed; et al	La-Z-Boy®	1,789,337	1931
Kraft, John	Processed Cheese	2,641,545	1953
Lake, Simon	Submarine	581,213	1897
Land, Edwin; et al	Instant Film	2,543,181	1951
Lawrence, Ernest	Atom Splitting Cyclotron	1,948,384	1934
Lear, William	Car Radio	1,959,869	1934
Leder, Timothy; et al	Genetically-Modified Animal	4736866	1988
Lincoln, Abraham	Buoying Device	6,469	1849
Loud, John	Ballpoint Pen	392,046	1888
Magee, Carl	Parking Meter	2,118,318	1938
Marconi, Guglielmo	Wireless Telegraphy	586,193	1897
Markey, Hedy	Secret Encoding System	2,292,387	1942
Mason, John	Mason Jar	22,186	1858
McCormick, Cyrus	Wheat Reaper	X8277	1834
McCoy, Elijah	Oil Lubricator	139,407	1873
McIntyre, Frank	Lead Pencil Design	D37431	1905
Mergenthaler, Ottmar	Linotype® Machine	304,272	1884
Morgan, Garrett	Traffic Signal	1,475,024	1923
Morse, Samuel	Telegraph	1,647	1840
Nobel, Alfred	Dynamite	57,175	1866
Noyce, Robert	Integrated Circuit	2,981,877	1961
Obara, Hiroyuki	Reconfigurable Toy	4,516,948	1985
Osius, Fred	Waring Blendor®	2,109,501	1938
Otis, Elisha	Elevator	31,128	1861
Otto, Nikolaus	Gasoline-Motor Engine	194,047	1877
Owens, Michael	Glass Bottle Machine	766,768	1904

Index

INVENTOR	INVENTION	PATENT	DATE
Painter, William	Bottle Cap	468,226	1892
Pasteur, Louis	Pasteurization	135,245	1873
Pope, Augustus	Burglar Alarm	9,802	1853
Rockola, David	Jukebox	3,129,005	1964
Roebling, John A.	Steel Cabling	2,720	1842
Russell, Lillian	Dresser-Trunk	1,014,853	1912
Ryan, John; et al	Barbie®	3,425,155	1969
Saunders, Clarence	Piggly Wiggly® Self-Serve Store	1,242,872	1917
Schawlow, Arthur	Optical Maser	2,929,922	1960
Schooley, Matthew	Paper Clip	601,384	1898
Selden, George	Horseless Carriage	549,160	1895
Shockley, William	Transistor	2,569,347	1951
Sholes, Christopher; et al	Typewriter	79,265	1868
Sikorsky, Igor	Helicopter	1,848,389	1932
Singer, Issac	Sewing Machine	13,661	1855
Smith, Horace; et al	Fire-Arm	10,535	1854
Spalding, Reuben	Winged Man	398,984	1889
Speers, Samuel; et al	G.I. Joe®	3,277,602	1966
Spencer, Percy	Microwave Oven	2,495,429	1953
Stevens, Carlile; et al	Clapper®	5,493,618	1996
Stickely, Gustav	Mission-Style Furniture	D37508	1905
Stillson, Daniel	Monkey Wrench	107,304	1870
Stone, Marvin	Paper Drinking Straw	375,962	1888
Struble, Glenn	French Fry Holder	3,630,430	1971
Stuart, James E.B.	Saber Attachment	25,684	1859
Taylor, Frederick	Golf Putter	792,631	1905
Tesla, Nikola	Electromagnetic Motor	381,968	1888
Vestor, Franz	Coffin Alert	81,437	1868
Westinghouse, George	Railroad Air Brakes	88,929	1869
Whitney, Eli	Cotton Gin	X72	1794
Woodland, Norm	Barcode	2,612,994	1952
Wright, John Loyd	Lincoln Logs®	1,351,086	1920
Wright, Orville; et al	Airplane	821,393	1906
Wyeth, Nathaniel C; et al	Plastic Molding Process	3,935,358	1976
Yale, Linus Jr.	Keyed Lock	31,278	1861
Zamboni, Frank	Ice Rink Cleaner	2,642,679	1953
Zeppelin, Ferdinand Graf	Navigable Balloon	621,195	1899
Zworykin, Vladimir	Improved Television System	2,141,059	1938

The Patents...

US005318296A

United States Patent [19]

Adams et al.

[11] **Patent Number: 5,318,296**

[45] **Date of Patent: Jun. 7, 1994**

[54] **MATCHED SETS FOR GOLF CLUBS HAVING MAXIMUM EFFECTIVE MOMENT OF INERTIA**

[75] Inventors: **Byron H. Adams**, Dallas, Tex.; **Faris W. McMullin**, Boise, Id.

[73] Assignee: **Adams Golf Inc.**, Richardson, Tex.

[21] Appl. No.: **974,707**

[22] Filed: **Nov. 12, 1992**

[51] **Int. Cl.**[5] **A63B 53/00**

[52] **U.S. Cl.** .... **273/77 A**; 73/65.03; 273/80 A

[58] **Field of Search** .... 273/77 R, 77 A, 80 A, 273/80 R, 80 C, 167 G; 73/65

[56] **References Cited**

U.S. PATENT DOCUMENTS

3,473,370 10/1969 Marciniak 73/65
3,698,239 10/1972 Everett, III 73/65
3,703,824 11/1972 Osborne et al. 73/65
4,203,598 5/1980 Stuff et al. .... 273/80 A
4,212,193 7/1980 Turley 73/65
4,240,631 12/1980 MacDougall .... 273/77 A
4,420,156 12/1983 Campau .... 273/77 A
4,674,324 6/1987 Benoit .... 273/77 A
4,840,380 6/1989 Kajita et al. .... 273/77 A
4,971,321 11/1990 Davis .... 273/77 A
5,094,101 3/1992 Chastonay .... 273/77 A

Primary Examiner—Vincent Millin
Assistant Examiner—Steven B. Wong
Attorney, Agent, or Firm—Aquilino & Welsh

[57] **ABSTRACT**

A dynamically matched set of golf clubs having a variable optimum moment of inertia for each club in the set wherein each club has a moment of inertia proportional to the lie angle of the club and the optimum value of the moment of inertia is determined using precise mathematical formulas and calculations.

13 Claims, 1 Drawing Sheet

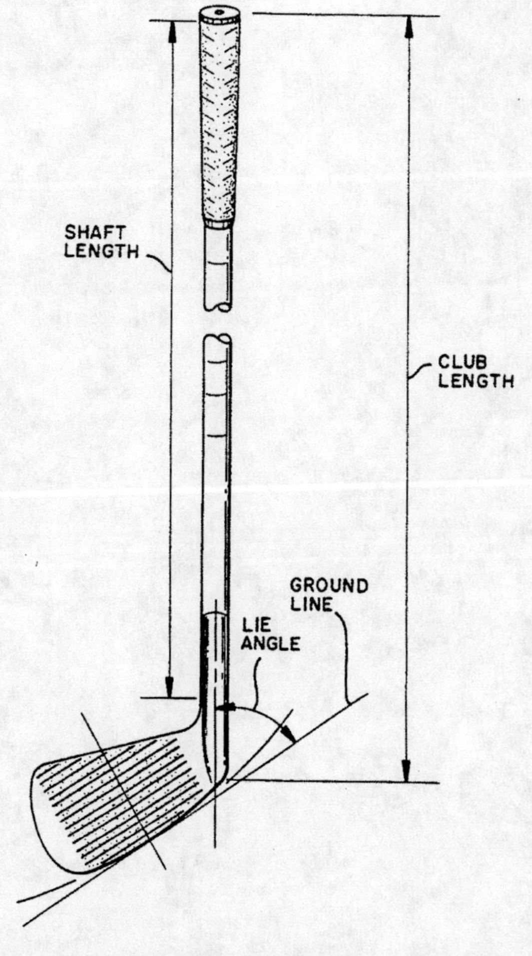

WOWidea! Technology rescues weekend hackers! Byron Adams found that when the sweet spot, where the clubface strikes the ball, is uniquely matched to the angle of the club head, the ball can be more properly put aloft. It simply proves that golfers, all over the world, love to find forgiveness no matter where their faults may lie.

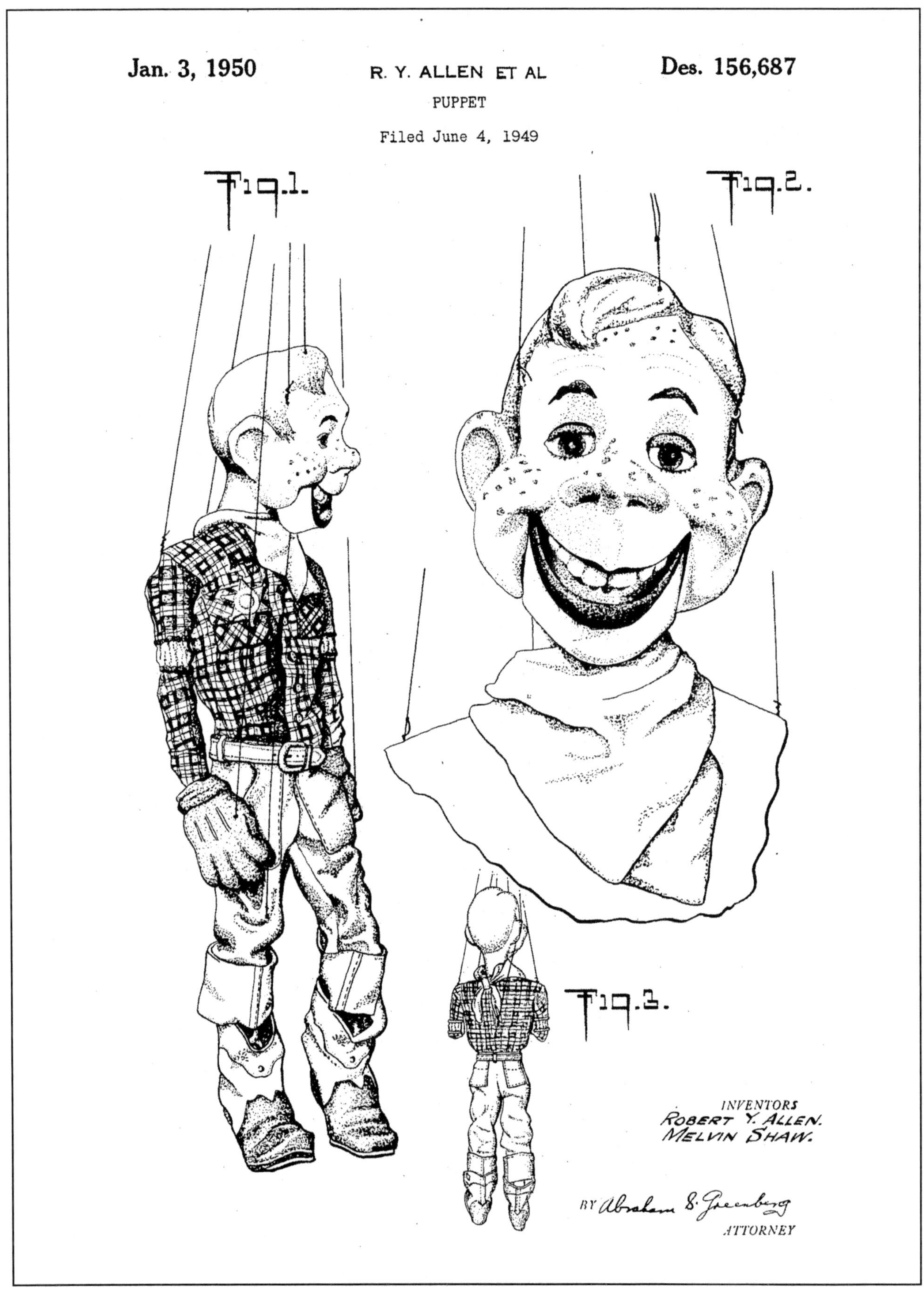

WOWidea! A toothy, grinning, freckle-faced marionette can win the hearts of real boys and girls! Robert Allen, who's Howdy Doody show brought laughter into millions of homes through his television series, also understood the more serious side of growing up, the need to share your awkward and gawky moments with a "friend".

Dec. 26, 1933. E. H. ARMSTRONG 1,941,066

RADIO SIGNALING SYSTEM

Filed July 30, 1930 2 Sheets-Sheet 1

Fig. 1.

INVENTOR
Edwin H. Armstrong.
BY Moses & Nolte
ATTORNEYS

WOWidea! Static-free radio! Until Edwin Armstrong developed the alternative, radio stations used amplitude modulation (AM) for broadcasting their programs, which tended to be affected by thunderstorms, electrical transmissions, and appliances. However, with his frequency modulation (FM) method, sweeter sounds were heard.

United States Patent [19]

Barnes et al.

[11] **3,782,612**

[45] **Jan. 1, 1974**

[54] **GARMENT HANGER**

[75] Inventors: **Dayton W. Barnes; Robert D. Chiles,** both of Greeneville, Tenn.

[73] Assignee: **Ja-San, Inc.,** Greeneville, Tenn.

[22] Filed: **July 11, 1972**

[21] Appl. No.: **270,672**

[52] **U.S. Cl.** **223/85**
[51] **Int. Cl.** **A41d 27/22**
[58] **Field of Search** 223/88, 98, 85

[56] **References Cited**

UNITED STATES PATENTS

2,260,607	10/1941	Coney	223/88
2,011,265	8/1935	Young	223/88
2,221,508	11/1940	Coney	223/88

Primary Examiner—Geo. V. Larkin
Attorney—J. Gibson Semmes

[57] **ABSTRACT**

A strut-type garment hanger is disclosed comprising two different length strands of wire or similar material, the strands being entwined and biased to create a longitudinal flexure within a strand holding tube and the overall hanger to give rigidity.

8 Claims, 4 Drawing Figures

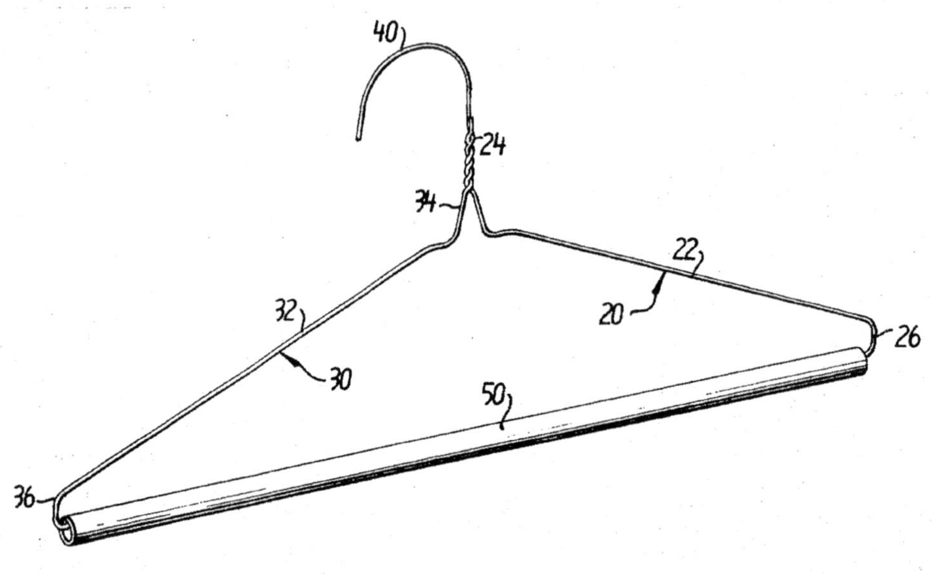

WOWidea! A new and improved clothes hanger! Barnes and Chiles' design reduced creases and costs while simultaneously increasing customer satisfaction. It just go to show that even the lowly hanger, made from two simple materials, can get an inventive work-over centuries after it's original arrival on the medieval clothes hook.

DESIGN.

A. BARTHOLDI.

Statue.

No. 11,023. Patented Feb. 18, 1879.

LIBERTY ENLIGHTENING THE WORLD.

E. A. Dick
J. E. Carpenter.

Auguste Bartholdi
By
A. Pollok
Atty.

WOWidea! A magnificent gift from France makes the concept of freedom tangible! Auguste Bartholdi's wondrous creation stands at the entrance to New York City's harbor. With its tower-like steel interior created by Gustave Eiffel, the Statue of Liberty, as it is better known today, beckons with her outstretched arm.

May 20, 1958 R. C. BAUMANN **2,835,548**

SATELLITE STRUCTURE

Filed Aug. 1, 1957 3 Sheets-Sheet 1

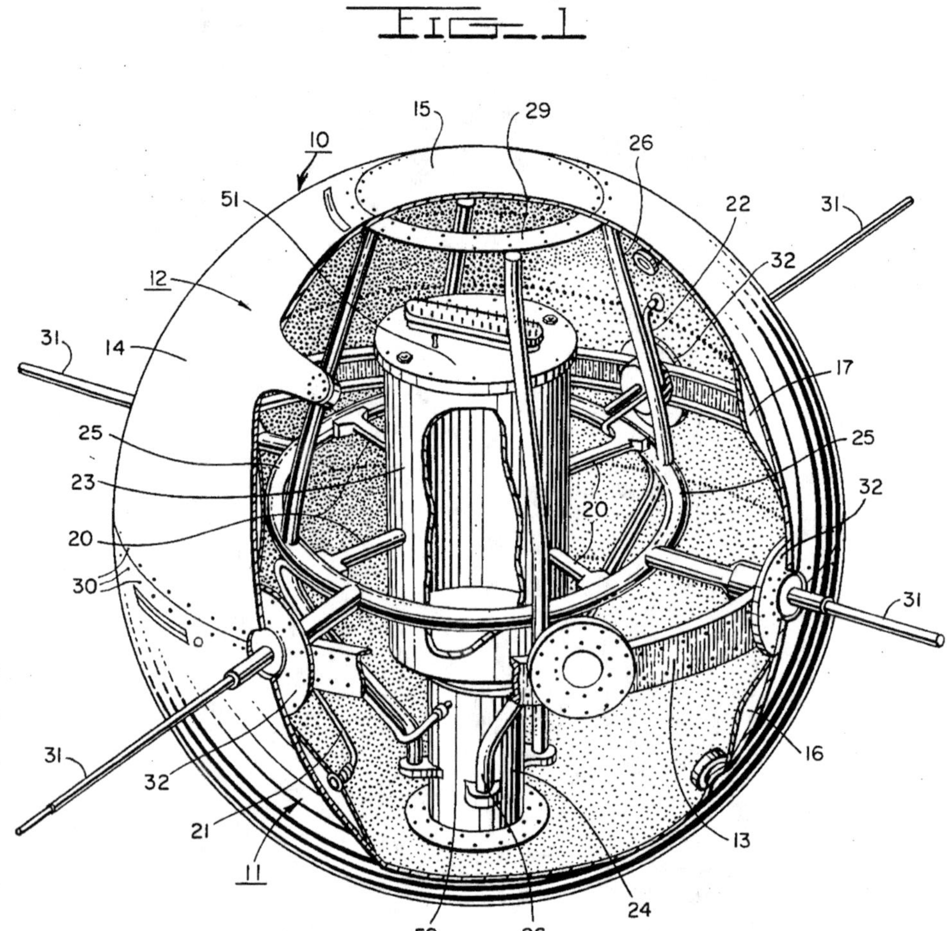

INVENTOR
ROBERT C. BAUMANN

BY W. R. Mattox
Richard C. Reed

ATTORNEYS

WOWidea! Calls to distant relatives can be out of this world! Launched March 17, 1958, less than 6 months after Russia's Sputnik I, Baumann's structural design of Vanguard I set the stage for early satellites providing wireless communications via geosynchronous orbit, a concept first put forth by science fiction writer Arthur C. Clarke.

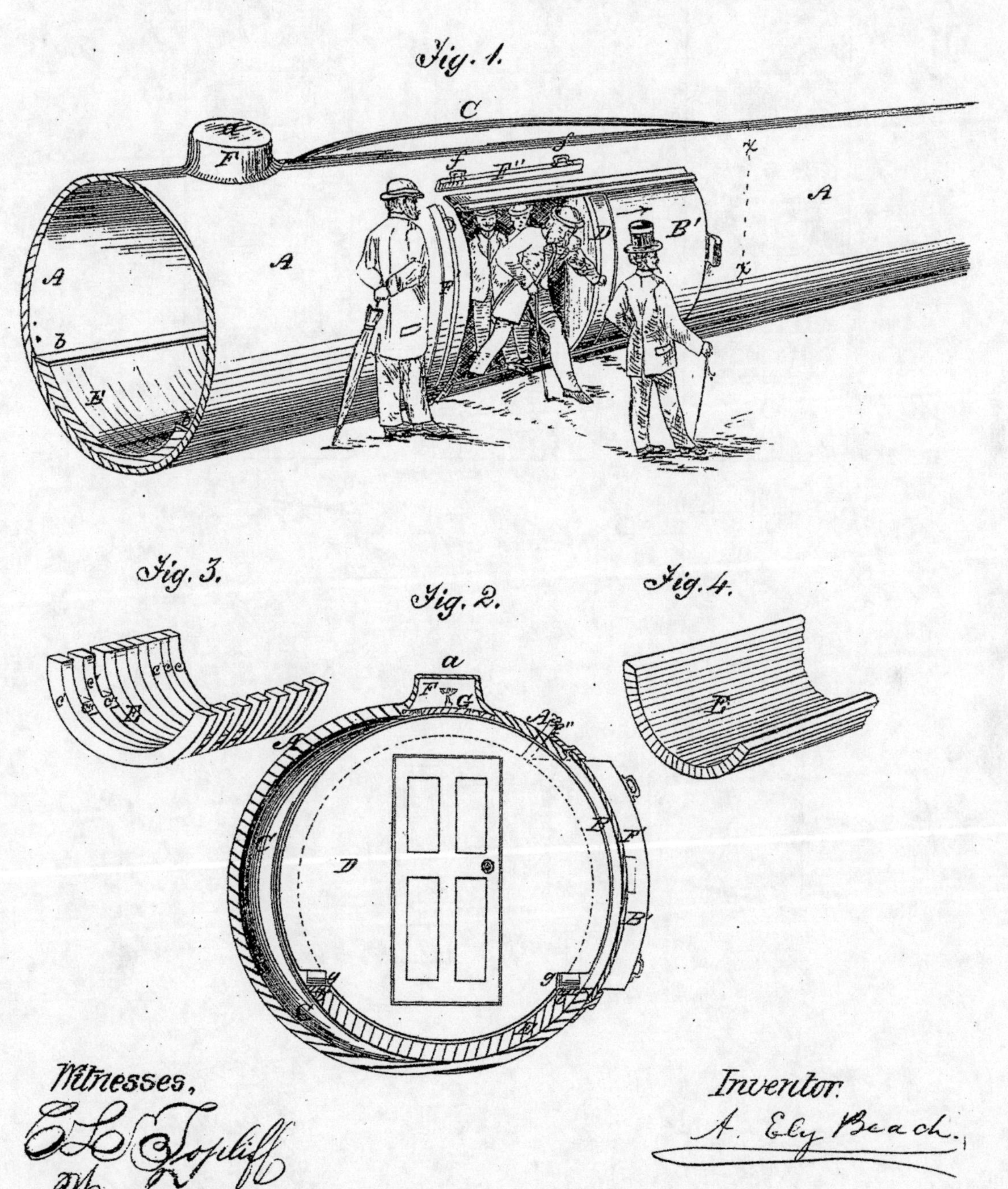

WOWidea! Build it and they will ride! Alfred Ely Beach was editor of *Scientific American* and, a big believer in new technology. As such, he asked for permission from New York City Hall to construct this air powered subway and they turned him down. Undaunted, he skirted the rules and finally had a demonstration line running in 58 days.

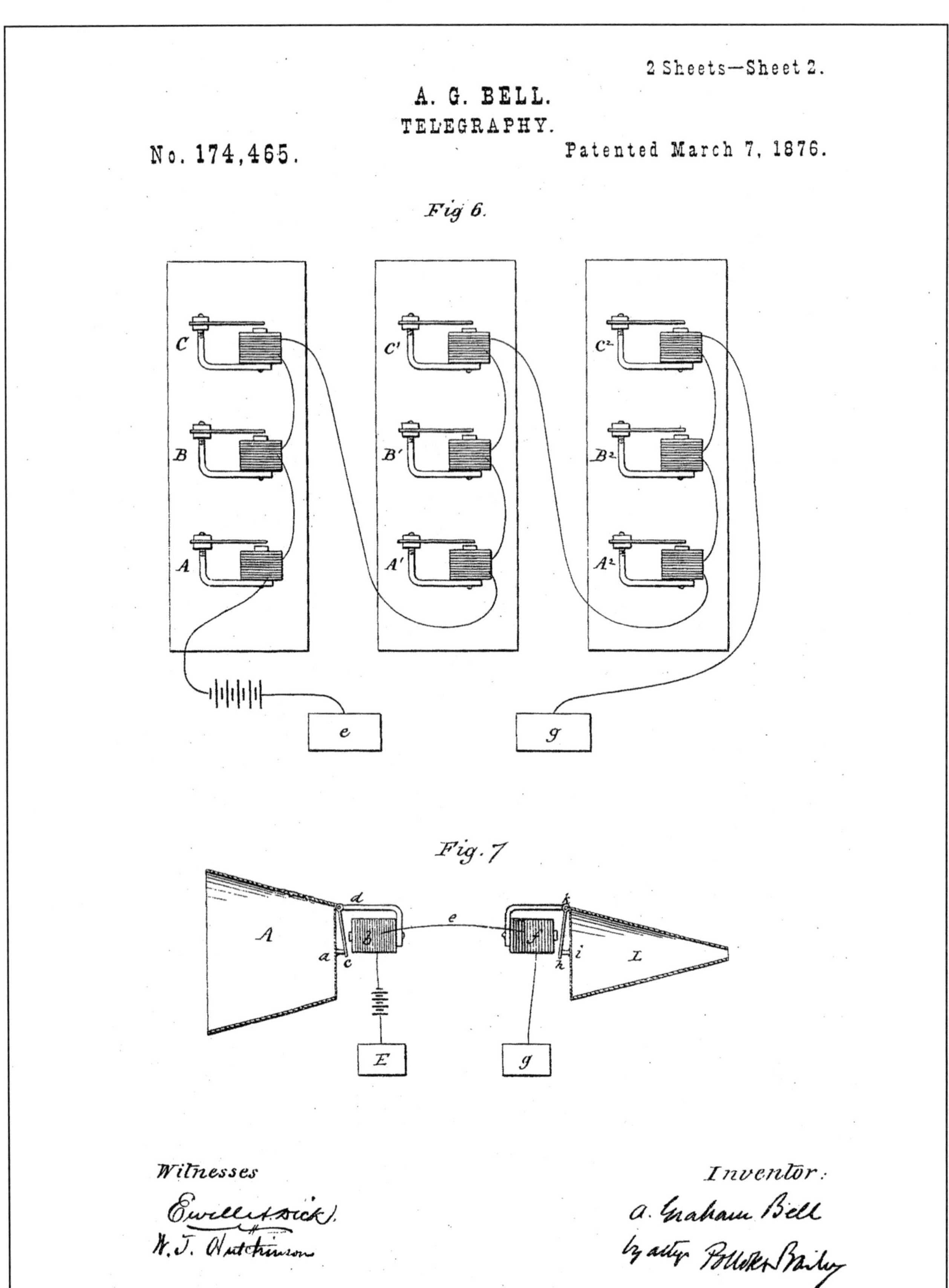

WOWidea! Teacher of the deaf invents a speech transmitting system! Underwritten by his future father-in-law, Alexander Graham Bell, with help from, Thomas Watson, Jr., beat their rival, Elisha Grey, to the Patent Office by hours, securing what many consider to be the most valuable United States patent ever issued--- the telephone!

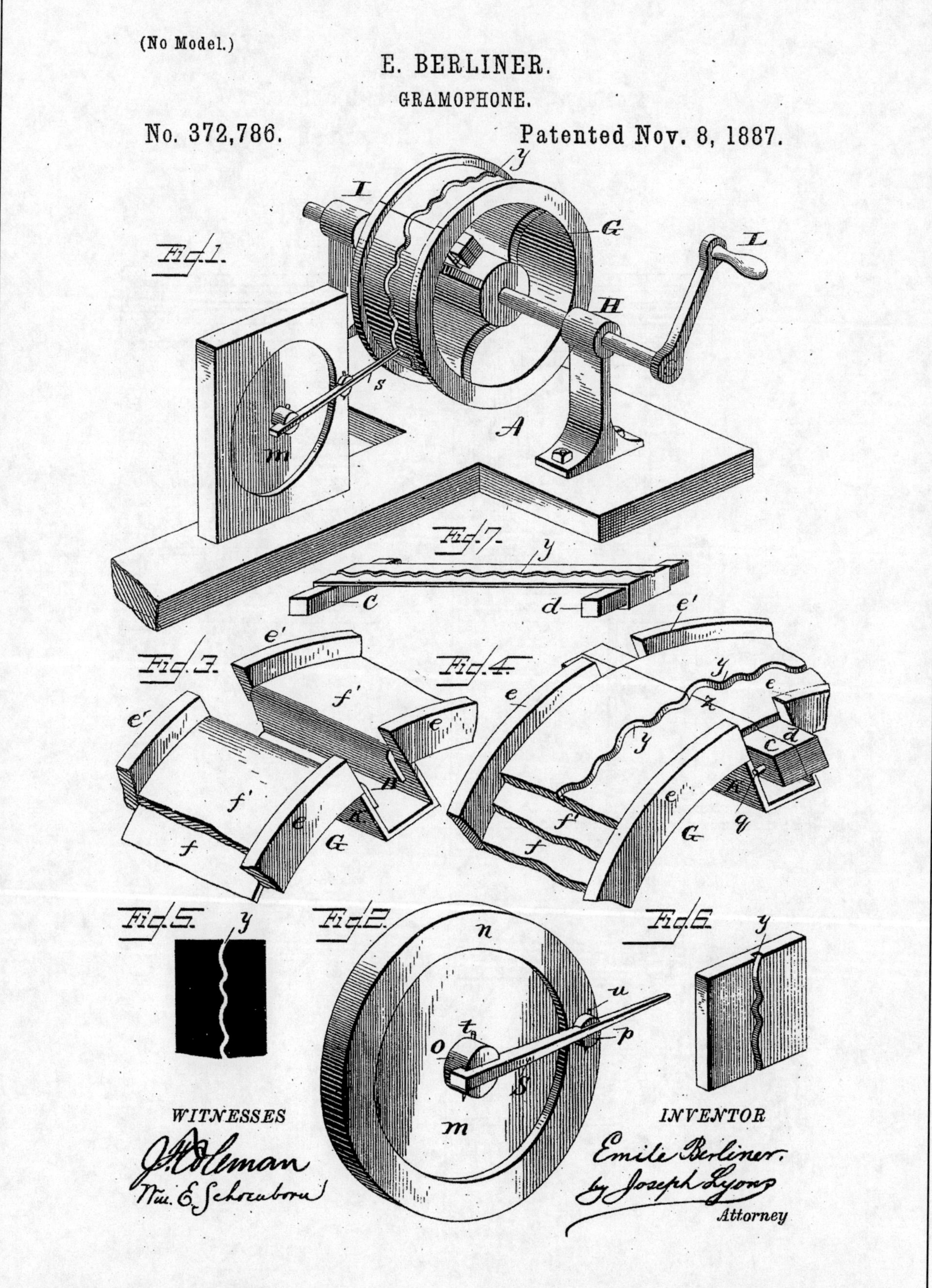

WOWidea! Side to side worked better than up and down! Owing much to Edison's pioneering efforts in sound reproduction, Emile Berliner's Gramophone makes long playing "records" a hit among audiophiles. A microphone inventor and former employee of Alexander Bell, he went on to create the famous Victor Talking Machine Co.

4 Sheets—Sheet 1.

H. BESSEMER.

MANUFACTURE OF IRON AND STEEL.

No. 16,082. Patented Nov. 11, 1856.

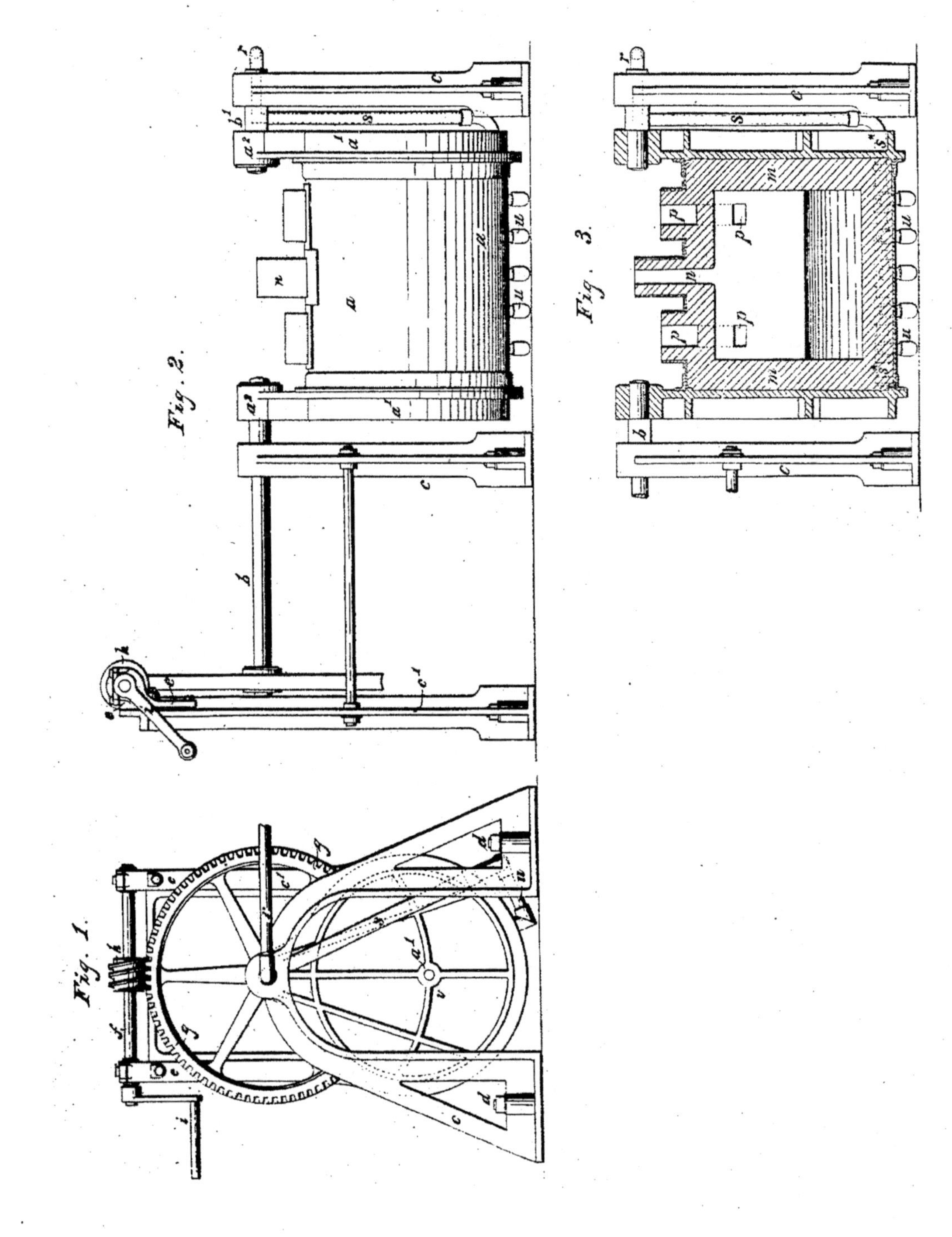

WOWidea! 1% is all it takes! Controlling carbon content makes it steel instead of weaker wrought or pig iron. British-born Bessemer conceived his invention in England, while Pennsylvania-based William Kelly independently patented a similar idea in the United States… the two concepts were eventually combined to perfect the process.

Aug. 12, 1930. C. BIRDSEYE **1,773,079**

METHOD OF PREPARING FOOD PRODUCTS

Filed June 18, 1927 5 Sheets-Sheet 5

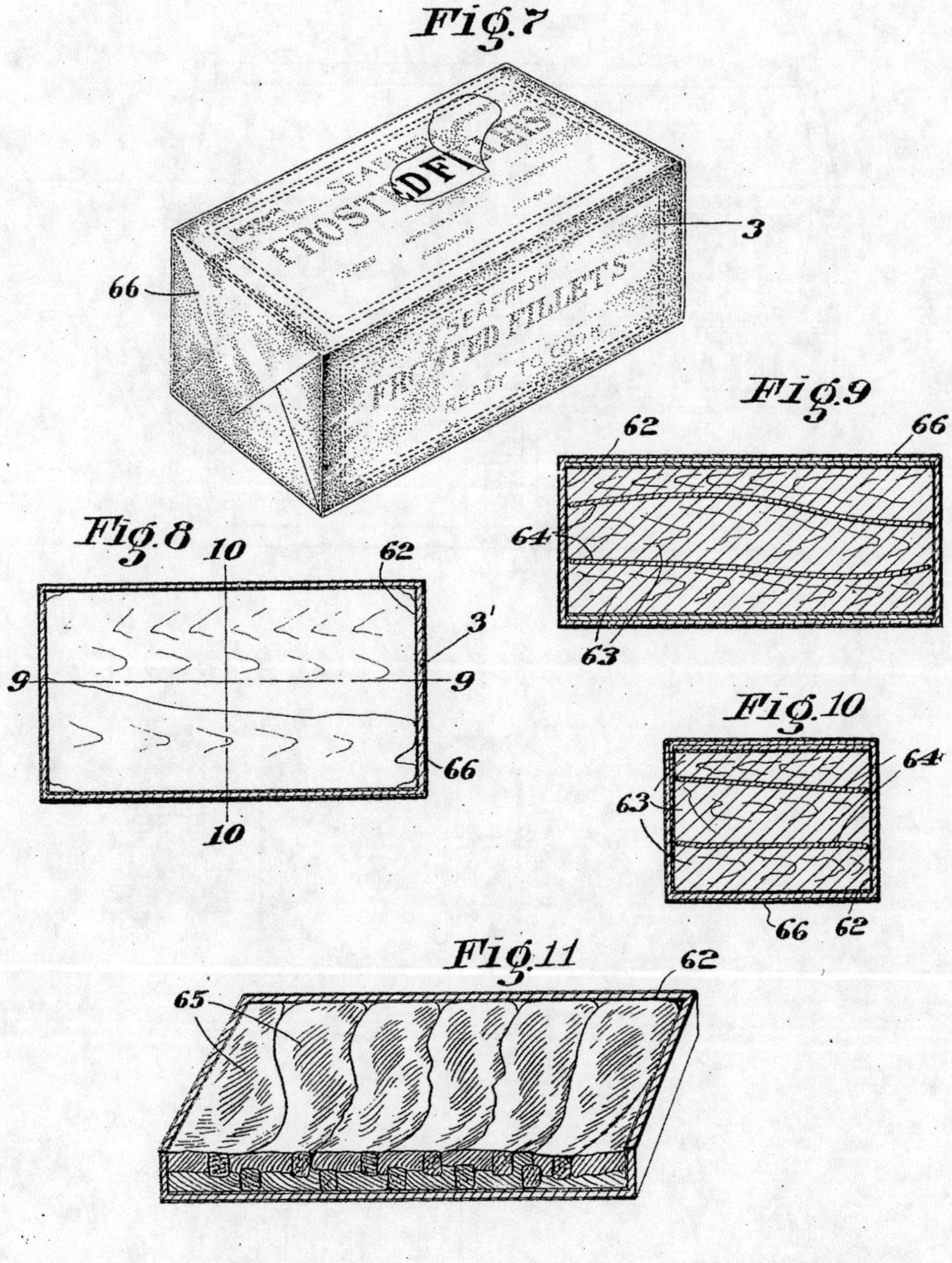

INVENTOR:
Clarence Birdseye
BY Hector M. Holmes
ATTORNEY

WOWidea! Naturalist moves to Labrador only to discover the secret of flash-freezing! Clarence Birdseye, who's patents and trademark would eventually create the General Foods Corporation, found that the arctic cold, when properly applied, could lock in a food's freshness, thereby allowing him to successfully market frozen haddock filets.

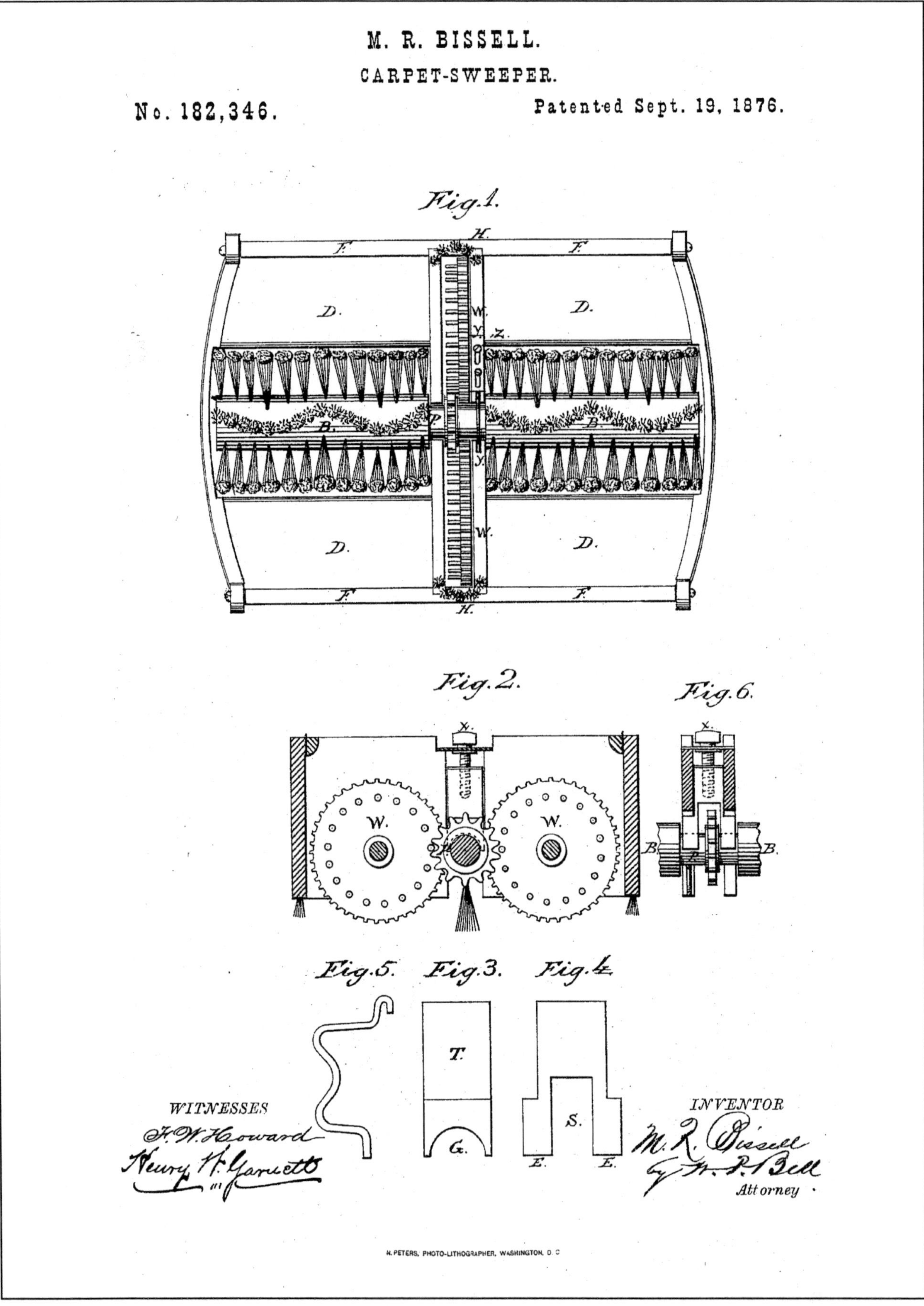

WOWidea! Combine the dust-pan and broom! Taking the rugs outside to beat them with a stick was "old hat" to Melvin Bissell. He had long believed that dust was the "root of all evil", and yet the process of keeping a home, or china shop like his clean, was arduous and time-consuming. His simply designed carpet sweepers swept the nation.

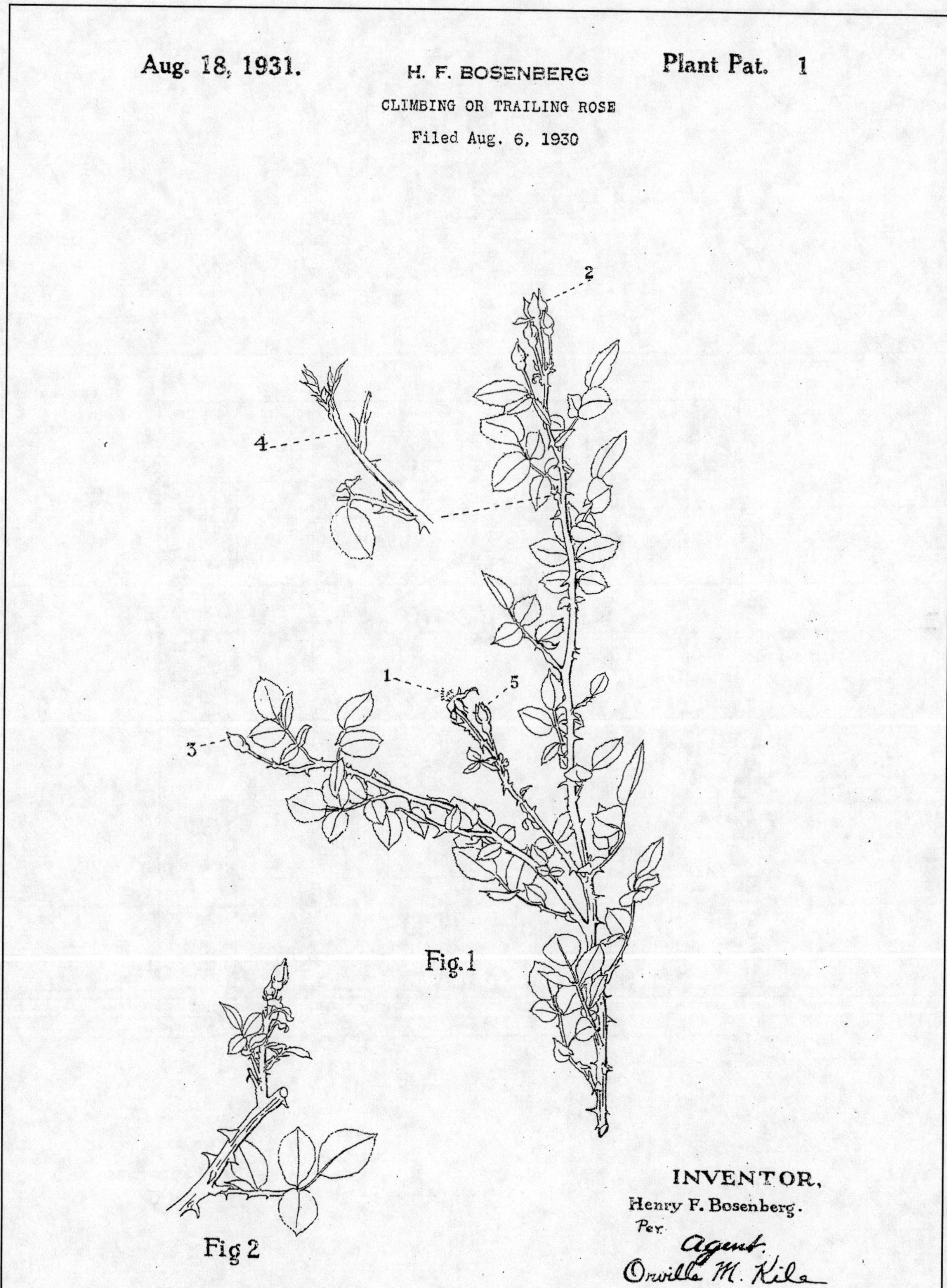

Aug. 18, 1931. H. F. BOSENBERG Plant Pat. 1

CLIMBING OR TRAILING ROSE

Filed Aug. 6, 1930

2

4

1

5

3

Fig.1

Fig 2

INVENTOR,

Henry F. Bosenberg.

Per

agent.

Orville M. Kile

WOWidea! Landscape gardener from New Jersey grows observant! The first asexual plant patent to be awarded in the United States went to Henry Bosenberg for his work in grafting roses which he noticed tended to bloom for extended periods of time. He named his rose, New Dawn, having the unique characteristics of endless blossoms.

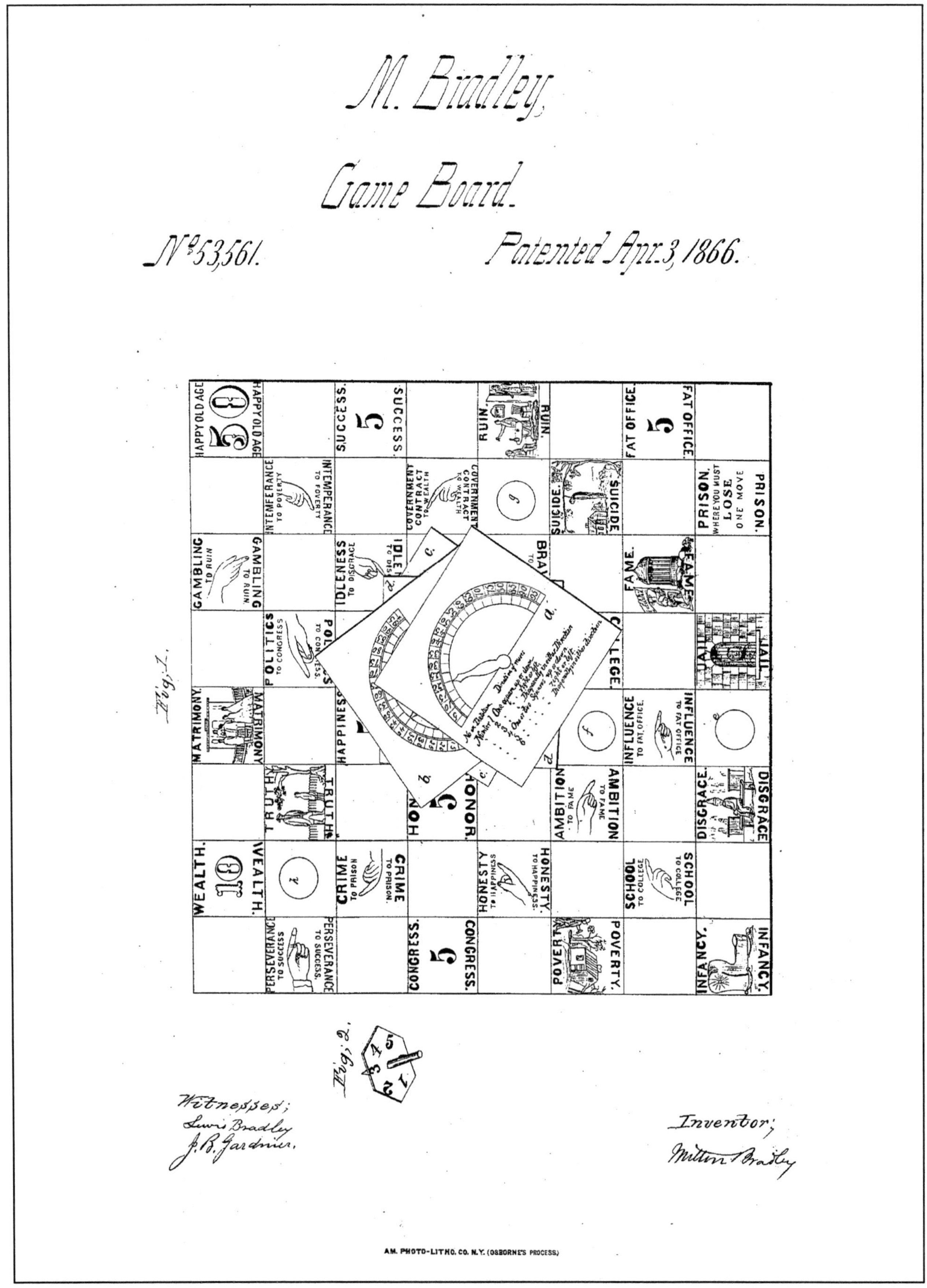

WOWidea! Teach life's twists and turns in one evening! Milton Bradley, a printer, nearly lost everything when he printed thousands of images of Abraham Lincoln as a clean-shaven candidate. When Lincoln grew his trademark beard, sales evaporated. Bradley then came up with what became The Game of Life®, a bestseller that saved his.

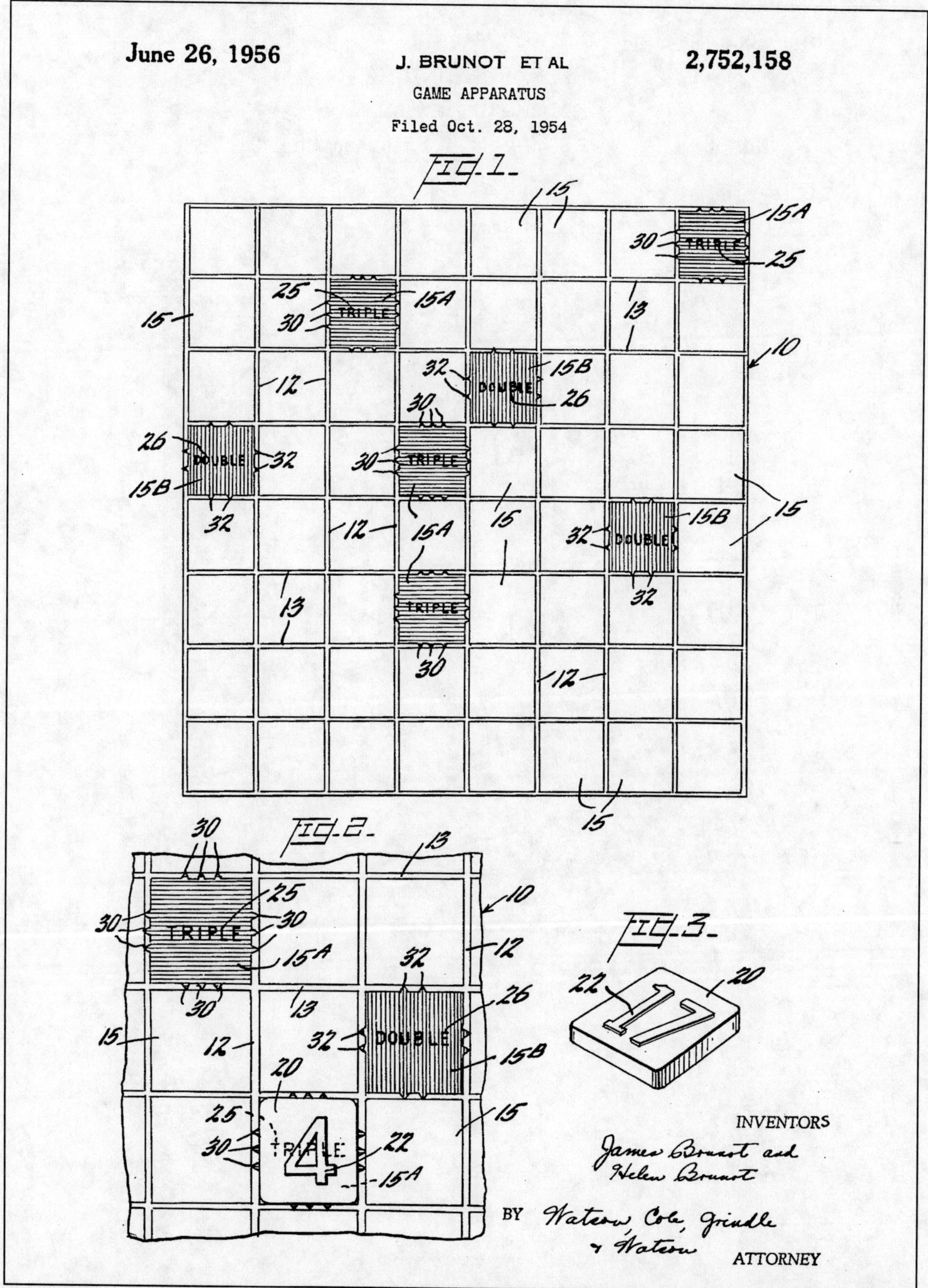

WOWidea! Giving up spelled success! Alfred Butts originated the word game he tried calling Lexico and Cris-Crosswords, but was unable to convince the Patent Office of its merits. In return for royalties, he agreed to let James and Helen Brunot give it a shot. They changed a few elements, secured a patent and renamed it Scrabble®.

(No Model.) 6 Sheets—Sheet 1.

W. S. BURROUGHS.

CALCULATING MACHINE.

No. 388,117. Patented Aug. 21, 1888.

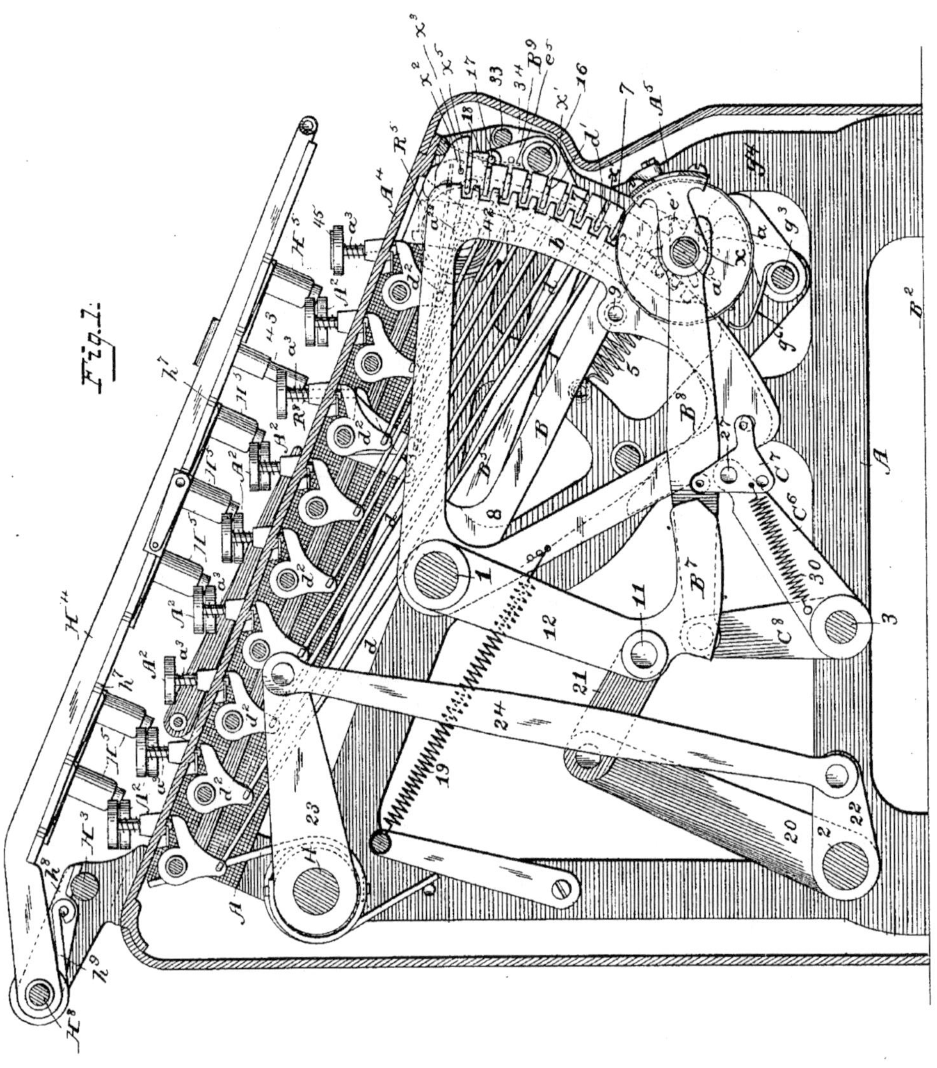

Attest:
Court A. Cooper.
K. E. Hansmann.

Wm. S. Burroughs.
Inventor:
By Foster & Freeman.
Attys.

N. PETERS, Photo-Lithographer, Washington, D. C.

WOWidea! A printer adds up to commercial success! Although an Asian invention, the abacus, predates William Burroughs by two thousand years, Burroughs was the innovator of a machine that not only calculated numbers but also printed a paper record. Initially prone to jam, he reworked his design and persevered to make its use foolproof.

United States Patent [19]

Bushnell

[11] **3,793,483**

[45] **Feb. 19, 1974**

[54] **VIDEO IMAGE POSITIONING CONTROL SYSTEM FOR AMUSEMENT DEVICE**

[76] Inventor: **Nolan K. Bushnell,** 3572 Gibson, Santa Clara, Calif. 95051

[22] Filed: **Nov. 24, 1972**

[21] Appl. No.: **309,268**

[52] **U.S. Cl.**.... **178/69.5 TV,** 178/5.8 R, 178/7.3 R, 178/7.5 D, 178/69.5 TV, 340/324 AD

[51] **Int. Cl.**......................... **H04n 3/22,** H04n 5/68

[58] **Field of Search**340/146 AE, 347 DA, 347 AD, 340/324 AD, 146.3 F, 168 R, 273, 324 A, 324 R; 273/DIG. 28; 178/6.8, DIG. 29, 5.8 R, 69.5 TV, 69.5 G, 6, 7.3 D, 7.3 S, 7.3 R

[56] **References Cited**

UNITED STATES PATENTS

3,659,285	4/1972	Baer	178/5.8 R
3,631,457	12/1971	Hamada	340/324 A
3,659,284	4/1972	Baer	178/5.8 R

Primary Examiner—Robert L. Richardson
Assistant Examiner—R. John Godfrey
Attorney, Agent, or Firm—Flehr, Hohback, Test, Albritton & Herbert

[57] **ABSTRACT**

For controlling the location of an image and to cause the image to move variously with respect to perpendicular coordinates, such as X, Y coordinates, on a video display tube, a first set of counters is arranged to generate artificial, horizontal and vertical sync pulses for use in conjunction with a video adder for controlling the image on a TV screen. A second set of counters driven from the same clock source as the first supplies information signals to the video adder for controlling the location at which the image will be displayed. Each of the two predetermined counters constituting the second set of counters is capable of being preset to any of a plurality of counts so as to cause a horizontal or vertical displacement of the image on the face of the display tube with respect to the locus defined by the count generated by the first set of counters.

5 Claims, 1 Drawing Figure

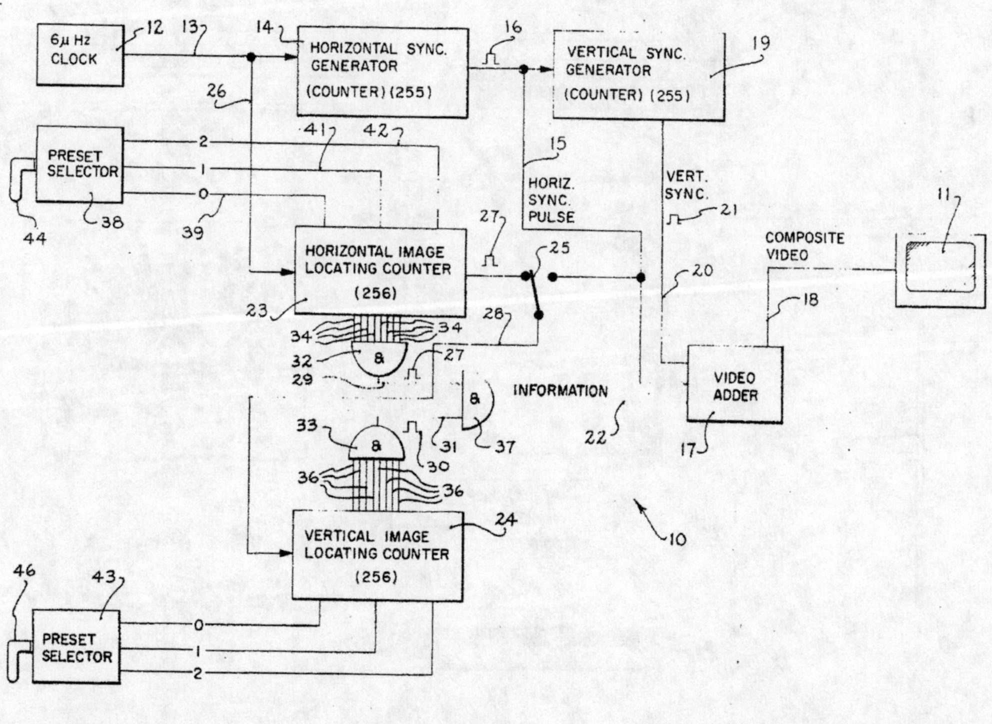

WOWidea! Who needs a table, net, ball, and paddles! Table tennis got an electronic cousin when Nolan Bushnell built one of the earliest forms of video games, Pong®. Played on a television set, it created hours of family fun. After creating Atari®, Bushnell went on to more consuming projects by founding Chuck E Cheese® Restaurants.

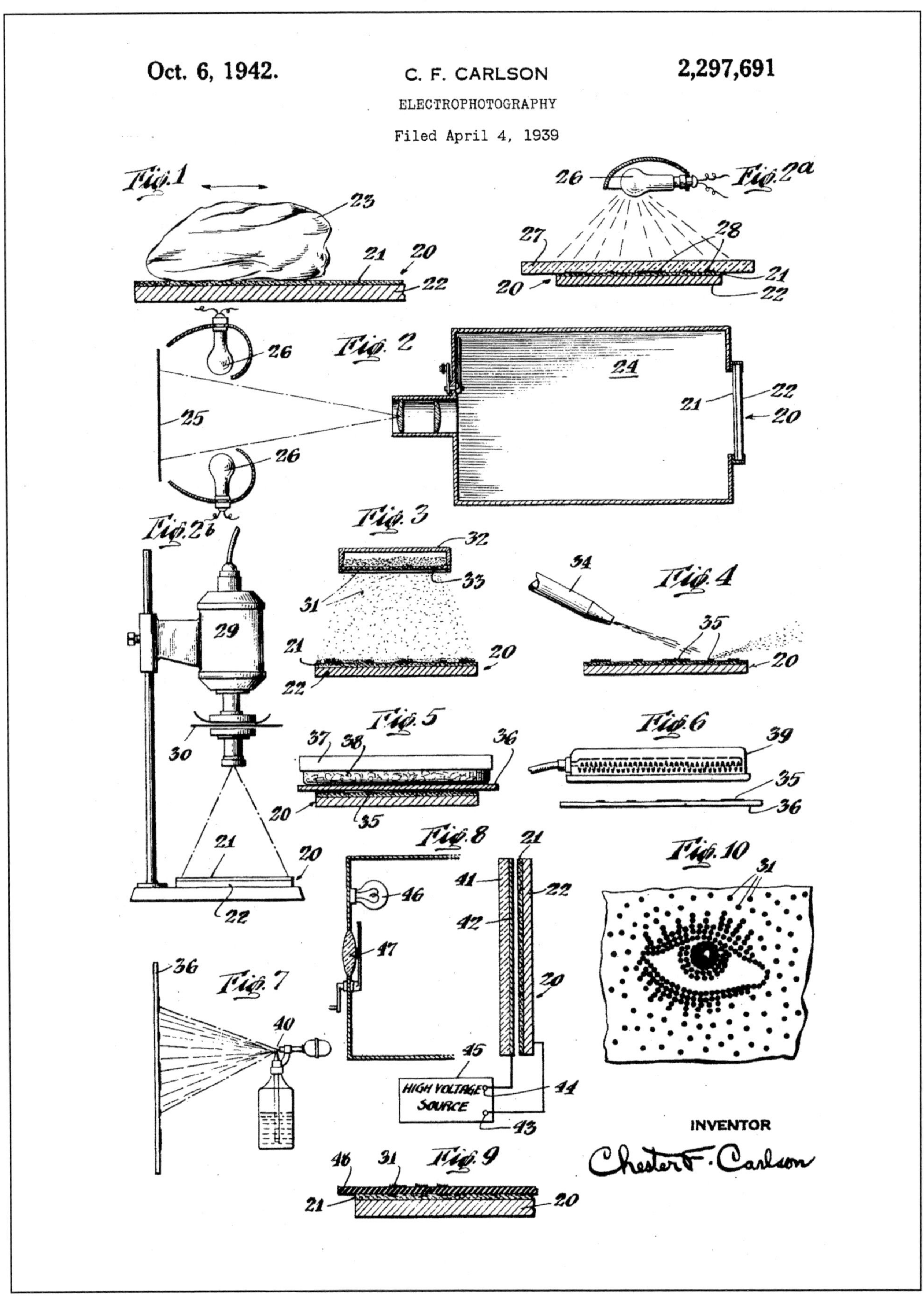

WOWidea! Months working in a patent library sheds light on solution! As a private law clerk, Chester Carlson had seen the need for a "copier" and was determined to find an answer. More than twenty firms turned him down before he found the Haloid Corp., which agreed to try his new invention, eventually changing its name to Xerox®.

Feb. 16, 1937. W. H. CAROTHERS 2,071,250

LINEAR CONDENSATION POLYMERS

Filed July 3, 1931

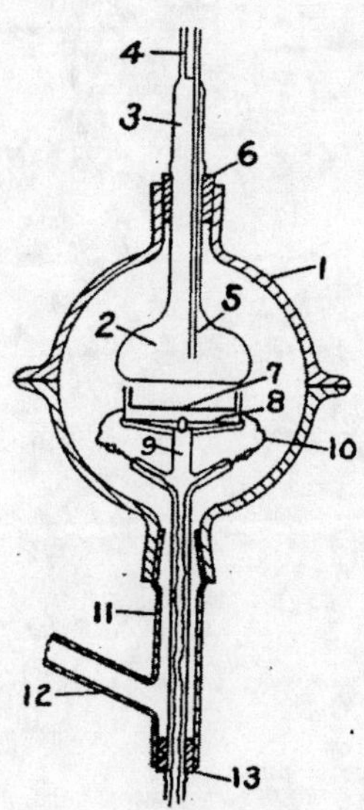

Wallace H. Carothers. Inventor

By R. F. Miller

Attorney

WOWidea! Some goop left overnight by accident is found to be both stretchy and strong! Wallace Carothers, while research team leader at DuPont, discovers the right combination of chemicals and processes to make the first totally synthetic fiber, Nylon®. First used in ladies stockings, sixty four million pairs were sold in the first year.

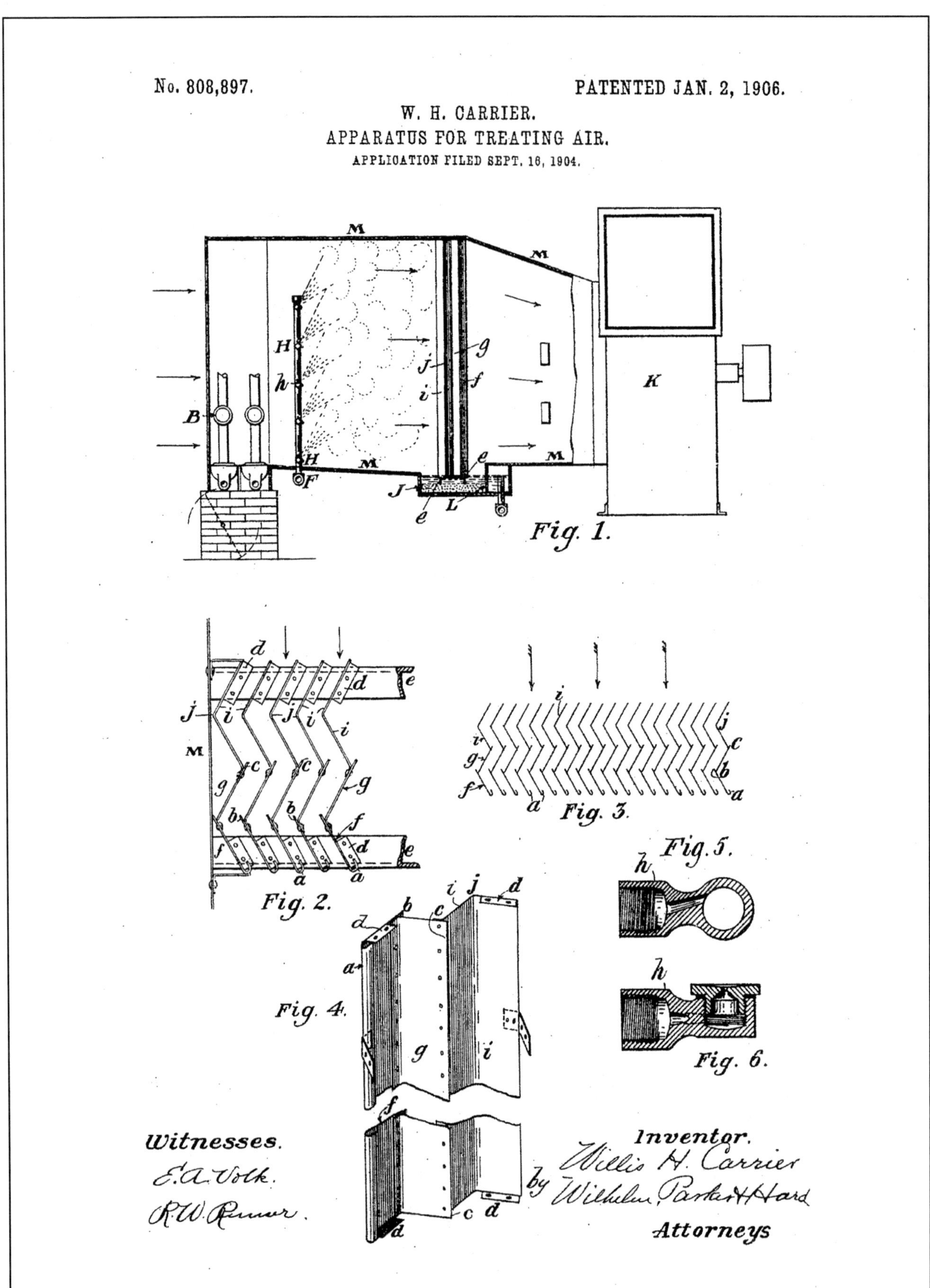

WOWidea! Understanding the relationship between temperature and humidity was the secret! Lack of money delayed his entry to college, but Willis Carrier ultimately became a money-saving engineer by conditioning air to solve costly industrial problems. Hot summers would never be the same or much fun without his cool inventions.

Patented Jan. 6, 1925. **1,522,176**

UNITED STATES PATENT OFFICE.

GEORGE WASHINGTON CARVER, OF TUSKEGEE, ALABAMA.

COSMETIC AND PROCESS OF PRODUCING THE SAME.

No Drawing. Application filed September 17, 1923. Serial No. 663,302.

To all whom it may concern:

Be it known that I, George Washington Carver, a citizen of the United States, residing at Tuskegee, in the county of Macon and State of Alabama, have invented certain new and useful Improvements in Cosmetics and Processes of Producing the Same, of which the following is a specification.

The invention relates to cosmetics and has as an object the provision of a pomade or cream made from peanuts. A further object of the invention is the provision of a process for making a pomade from peanuts which will provide a "vanishing cream" of any desired or usual tint, the pomade or cream having powder combined therewith. To carry out the process, the peanuts may be utilized in their raw, boiled or blanched condition and are first ground or macerated in any desired manner to the fineness of peanut butter. If for any reason a granular pomade is desired the grinding of the peanuts is carried out only to the extent necessary to give the character desired to the finished product. When ground to the fineness of peanut butter as suggested the resulting product will be a perfectly smooth substance.

To the ground or macerated nuts taking as a basis one ounce of peanuts there is next added 100 c. c. of pure water either hot or cold which is well stirred in with the ground nuts.

The resulting mixture is then strained through a piece of cheese cloth with gentle pressure and is put on the stove or water bath and evaporated until the oil becomes plainly visible on the surface.

The resulting product may be used unmodified in the subsequent steps or 2 c. c. of peanut oil may be added and the entire mass stirred until it becomes of the consistency of thick cream.

The material is then removed from the fire and approximately six grams of toilet powder such as kaolin, kaolinite, or china clay (preferably having slight fuller's earth properties) is added and the combined mass is thoroughly mixed until it becomes a thick heavy cream.

A quantity of salicylic acid substantially the size of a small pea, 10 drops of benzoin, and three or four drops of any desired perfume are then added. The mass thus obtained is finally ground or macerated until absolutely smooth, if the smooth product is desired, and the product is packed in porcelain, or glass containers.

If desired the above process may be modified by omitting either the added peanut oil, or the toilet powders, or both. By proper choice of the toilet powder any desired color may be given the product, from the dark brunette shades through the pinks, lavenders, to pure white.

I claim:—

1. The process of producing a cosmetic which comprises reducing peanuts to a finely divided condition, diluting the product with water reducing the mass to a consistency of thick cream and adding a preservative thereto.

2. The process of producing a cosmetic which comprises reducing peanuts to a finely divided condition, adding peanut oil and a preservative thereto and reducing the mass to a consistency of thick cream.

3. The process of producing a cosmetic which comprises reducing peanuts to a finely divided condition, diluting the product, adding toilet powder and a preservative thereto and reducing the mass to a consistency of thick cream.

4. The process of producing a cosmetic which comprises reducing peanuts to a finely divided condition, diluting with water, heating the mixture, adding peanut oil and a preservative and reducing the mass to a consistency of thick cream.

5. The process of producing cosmetics which comprises reducing peanuts to a finely divided condition, diluting the product, evaporating until oil appears upon the surface, adding peanut oil, toilet powder and a preservative.

6. The process of producing cosmetics which comprises reducing peanuts to a finely divided condition, diluting the product, straining the diluted mass, evaporating until oil appears upon the surface, adding peanut oil, stirring toilet powder into the mass, adding a preservative and a perfume and macerating until smooth.

7. The process of producing a cosmetic which comprises reducing peanuts to a finely divided condition, diluting with water, straining, evaporating until oil appears upon the surface, adding peanut oil, stirring

WOWidea! One among his 325 peanut creations! Born into slavery, George Washington Carver became the foremost expert on crops specifically designed to renew the cotton and tobacco fields of post-Civil War South. By 1940, peanuts were the South's biggest cash crop and farmers around the world had adopted his growing ideas.

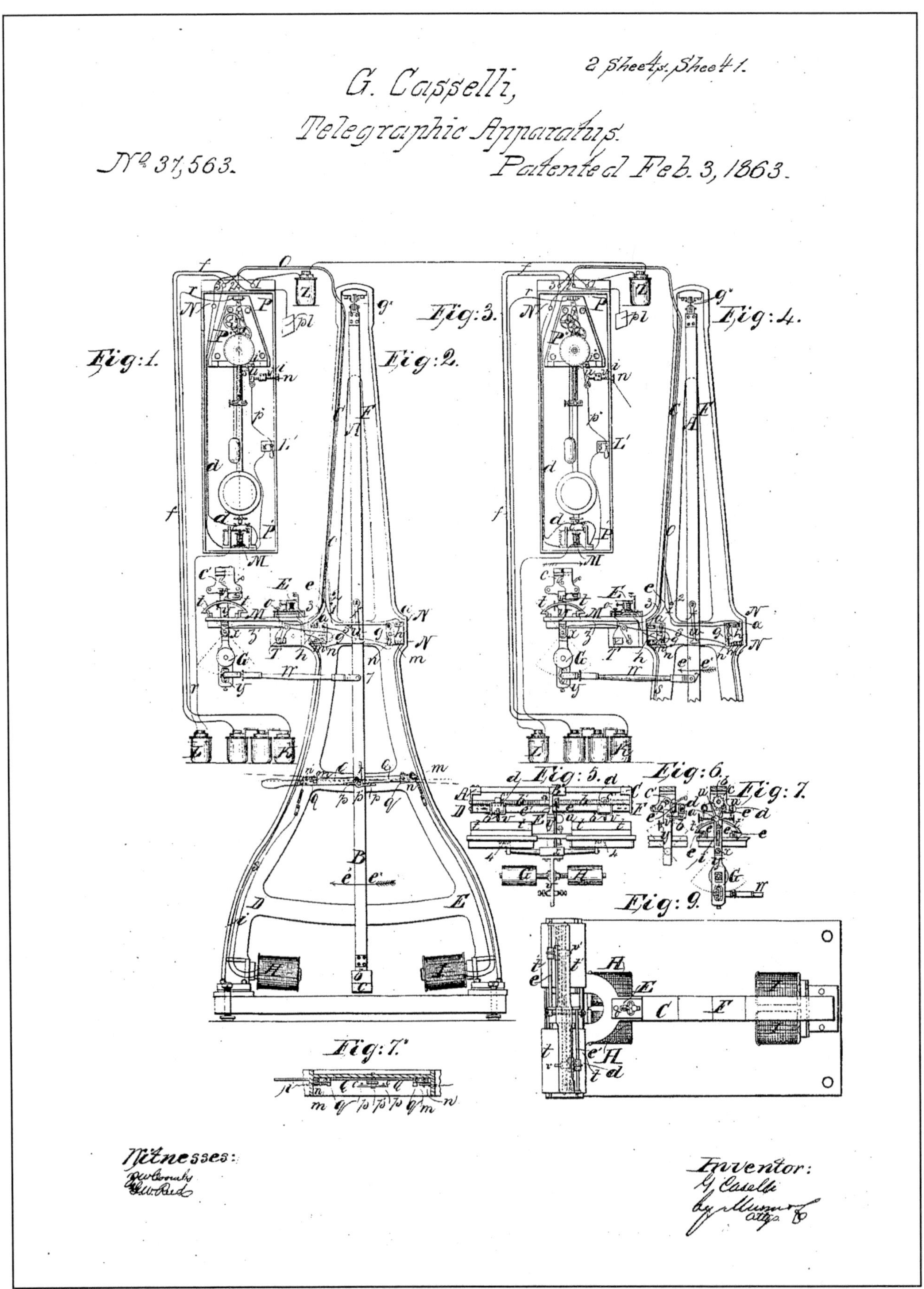

WOWidea! Italian priest drew upon the work of others to develop facsimile machine! Father Giovanni Caselli used existing telegraphic concepts to produce a pendulum that scanned specially treated drawings, recreating them miles away at a device which reversed the process. For a time, the French government used it to transmit military maps.

127,568

UNITED STATES PATENT OFFICE.

ROBERT A. CHESEBROUGH, OF NEW YORK, N. Y.

IMPROVEMENT IN PRODUCTS FROM PETROLEUM.

Specification forming part of Letters Patent No. 127,568, dated June 4, 1872.

To all whom it may concern:

Be it known that I, ROBERT A. CHESEBROUGH, of the city, county, and State of New York, have invented a new and useful Product from Petroleum, which I have named "Vaseline;" and I do hereby declare that the following is a full, clear, and exact description thereof, which will enable those skilled in the art to make and use the same.

The substance from which vaseline is made is the residuum of petroleum left in the still after the greater part of the petroleum has been distilled off. Distillation is now conducted in two ways—first, by applying to the still sufficient heat to vaporize all the oil therein down to the residuum; and second, by distilling under a vacuum, according to what is known as the "vacuum" process, by which less heat is required than in the ordinary way. The residuum left in the still by the vacuum process is greater in amount and better in quality than that produced by the old method, owing to its having been subjected to much less heat and to not having been run down so low in the still. The residuum produced by the vacuum process makes, therefore, a better quality of vaseline than that produced by the old process, and is more easily and cheaply refined by filtration through bone-black, but the products from both are identical in properties and appearance.

My method of making vaseline is by filtering the aforesaid petroleum residuums through bone-black, according to my process described in my Letters Patent dated August 22, 1865, and numbered 49,502, preferring to use in said process my steam-filter patented by me August 8, 1865, numbered 49,230, and preferring also to use bone-black, prepared as described in my patent dated July 10, 1866, numbered 56,179; although I do not confine myself to the use of the said improved filter or prepared bone-black.

Vaseline is the product of the filtration of the said residuums through bone-black, and varies in color as it comes from the filter. First it is a pure white at the beginning of the operation, soon changing to a light straw, then to a dark straw, and then to a deep claret at the close of the operation, the amounts of the several colors varying somewhat, according to the quality of the residuum, the quality, fineness, and preparation of the bone-black, and the skill of the manipulator in charging the filters and properly heating them.

Vaseline is a thick, oily, pasty substance; is semi-solid in appearance, unobjectionable in odor, becomes fluid at temperatures varying from 85° to 110° Fahrenheit, and, when fluid, is transparent. It will not saponify, does not crystallize, and does not contain paraffine, and in this respect essentially differs from the heavy products of petroleum which have been subjected to destructive distillation, and which are known as paraffine-oils.

When paraffine-oil is filtered by my process the result is a better article of paraffine-oil, and not vaseline. It will, therefore, be seen that vaseline cannot be made from any of the distilled products of petroleum, but only from the residuums of the still which have not been vaporized; and hence it is a distinct substance by itself, is a new article of manufacture, and is useful for various purposes.

Vaseline is especially useful in currying, stuffing, and oiling all kinds of leather. It is also a good lubricator, and may be used to great advantage on all kinds of machinery. The finest grade of vaseline is also adapted to use as a pomade for the hair, and will be found excellent for that purpose, one of its chief recommendations being that it does not oxidize. It is also an excellent substance for glycerine-cream for chapped hands, &c., which it resembles closely in touch and somewhat in appearance. Its gravity varies from 20° to 34° Baumé, according to the gravity of the crude vacuum or other residuum used in its production.

Claim.

What I claim as my invention, and desire to secure by Letters Patent, is—

The new article of manufacture named by me "Vaseline," substantially as herein described.

ROBT. A. CHESEBROUGH.

Witnesses:
CHAS. DRAKE,
JOHN TREDWELL.

WOWidea! Messy residue brings relief! Robert Chesebrough, a self-employed illuminating oil businessman, went to the hills of Pennsylvania, where oil was first discovered in America, to contract for supplies. Instead he hears of the magical medical properties of a oil-well byproduct. After years of self-inflicted wounds, petroleum jelly is born.

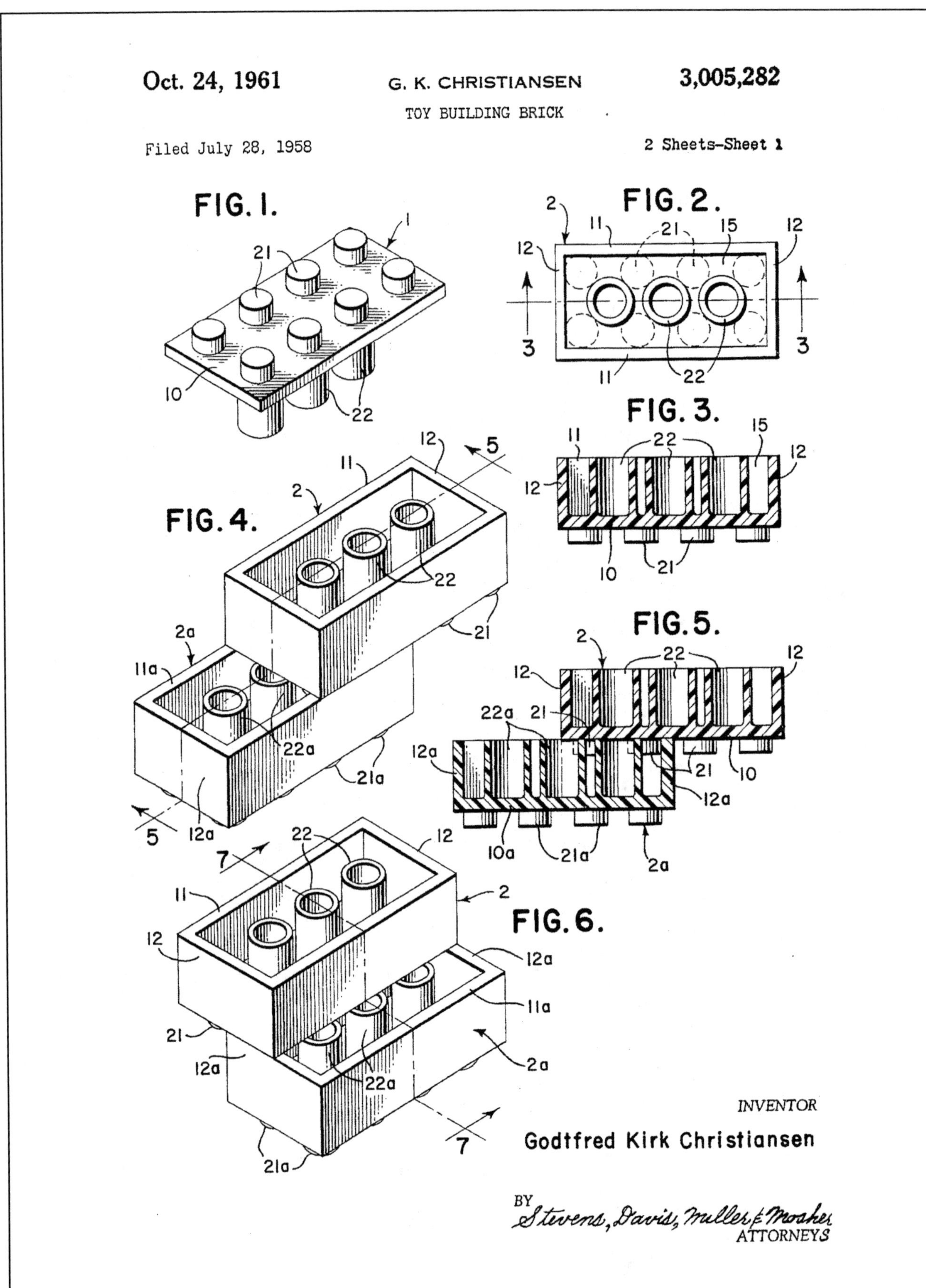

WOWidea! A line of toys that fit together! Godtfred Christiansen, son of a Danish wooden toy maker, listened to a customer's lament and envisioned a series of plastic building blocks that could be attached to one another to form most anything the mind could conceive. The word Lego® was created as Dutch shorthand for "play well".

(38.) SAMUEL L. CLEMENS.

Improvement in Adjustable and Detachable Straps for Garments.

No. 121,992. Patented Dec. 19, 1871.

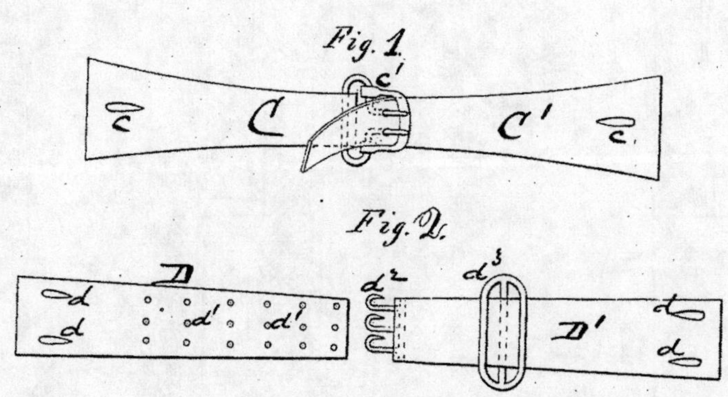

Witnesses:
Henry N. Miller
C. L. Evert.

Inventor
Saml. L. Clemens
per Alexander & Mason
Attorneys.

WOWidea! Strive to be more than meets the eye! Samuel Clemens certainly did do more than produce a style of men's suspenders. As Mark Twain, he authored many beloved and inventive tales, including "Tom Sawyer". Naturally, Clemens' creative streak led him to embrace other new ideas of his age, like the type writing machine.

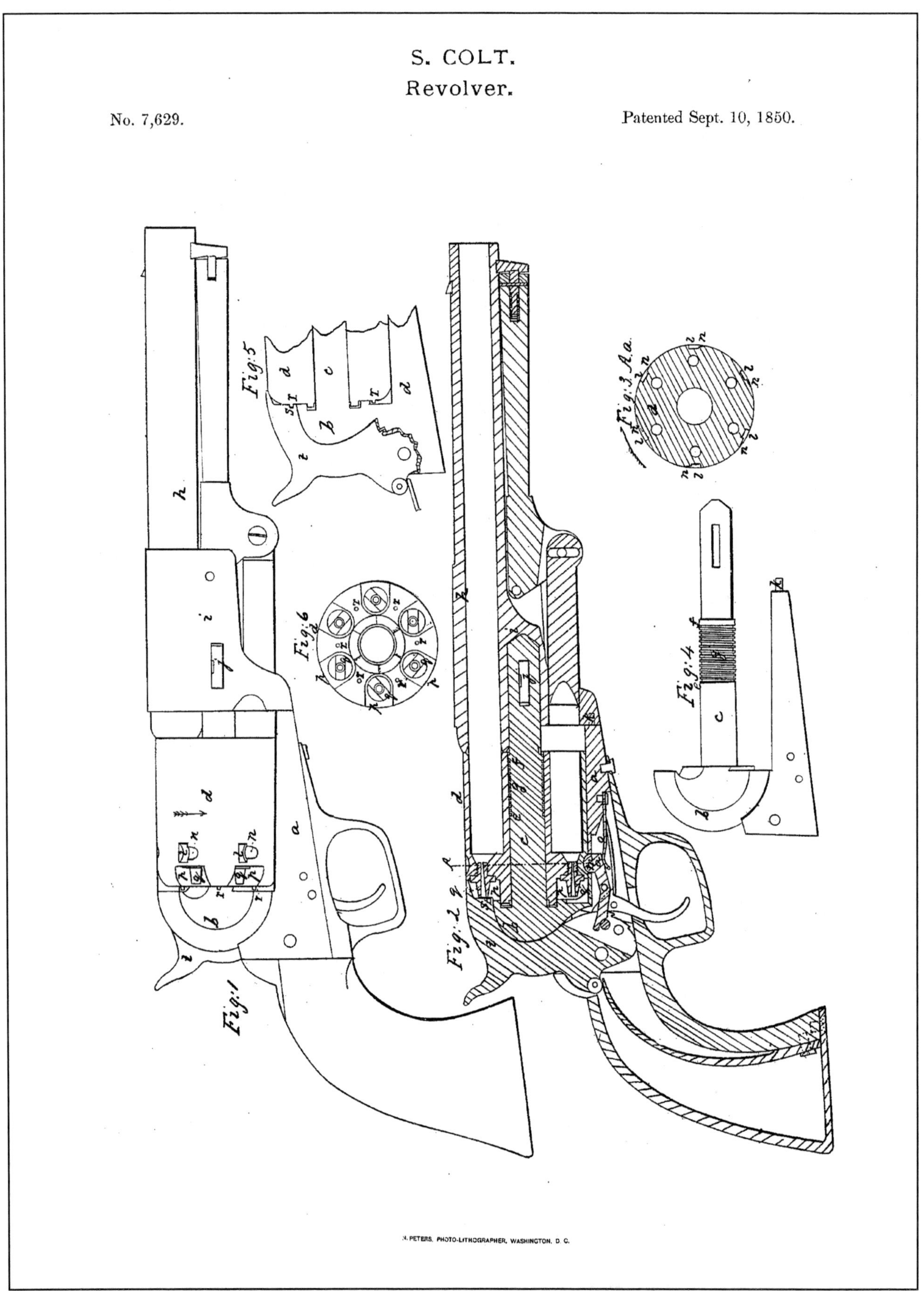

WOWidea! Why stop at failure! Samuel Colt invented his gun just as Texas won its independence from Mexico. So, without a war to fight, the Paterson, NJ-based company failed. He switched to the telegraph business to make ends meet. When conflicts returned, he enlisted Eli Whitney's help to successfully mass-produce guns.

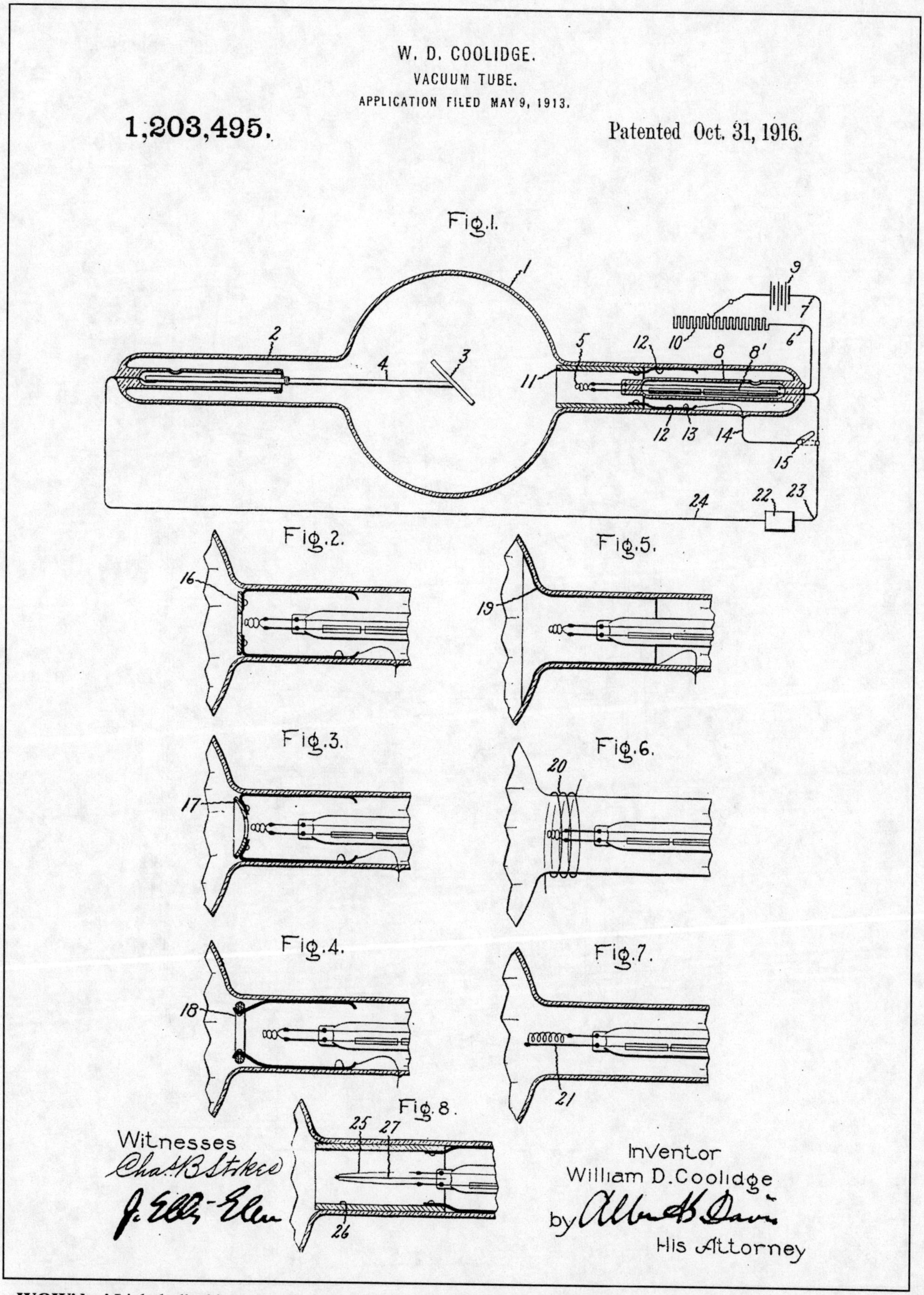

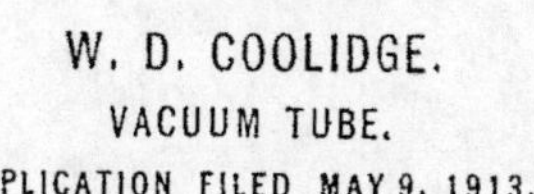
W. D. COOLIDGE.
VACUUM TUBE.
APPLICATION FILED MAY 9, 1913.

1,203,495. Patented Oct. 31, 1916.

Fig.1.

Fig.2.

Fig.5.

Fig.3.

Fig.6.

Fig.4.

Fig.7.

Fig.8.

Witnesses
Chas. B. Stikes
J. Ellis Elen

Inventor
William D. Coolidge
by Albert G. Davis
His Attorney

WOWidea! Light bulb shines on solution! William D. Coolidge, while in the employ of the General Electric Company, found that a certain element, tungsten, when specially treated, made the best filament for modern light bulbs. Taking tungsten a step further, inside a vacuum, he made the x-ray process safer and more convenient.

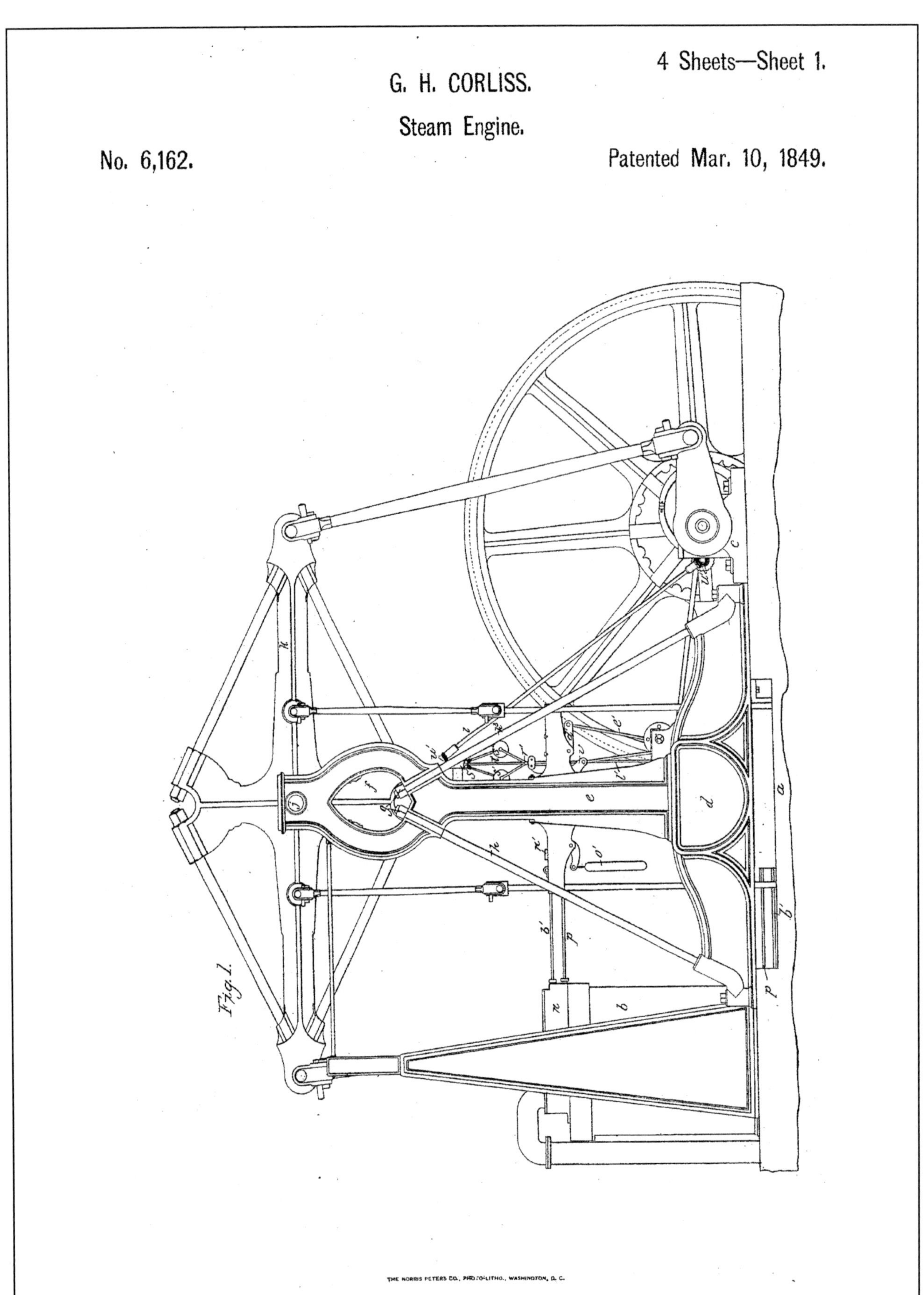

WOWidea! Constant speed meant money! James Watt is credited with the success of the steam engine, yet his designs could not adapt quickly enough when manufacturing processes suddenly required more or less power. Enter George Corliss, a shop keeper who had a penchant for mechanics and the answer to controlling steam valves.

United States Patent [19]

Damadian

[11] **3,789,832**

[45] **Feb. 5, 1974**

[54] **APPARATUS AND METHOD FOR DETECTING CANCER IN TISSUE**

[76] Inventor: **Raymond V. Damadian,** 64 Short Hill Rd., Forest Hill, N.Y. 11375

[22] Filed: **Mar. 17, 1972**

[21] Appl. No.: **235,624**

[52] **U.S. Cl.** .... **128/2 R,** 128/2 A, 324/.5 R
[51] **Int. Cl.** **A61b 5/05**
[58] **Field of Search** .... 128/2 R, 2 A, 1.3; 324/.5 A, 324/.5 B

[56] **References Cited**

UNITED STATES PATENTS

3,691,455	9/1972	Moisio et al.	324/.5 R
3,557,777	1/1971	Cohen	128/2 R
3,530,371	9/1970	Nelson et al.	324/.5 AC

OTHER PUBLICATIONS

Singer, J. R., Journ. of Applied Physics, Vol. 31, No. 1, Jan., 1960, pp. 125–127,

Primary Examiner—Kyle L. Howell
Attorney, Agent, or Firm—Brumbaugh, Graves, Donohue & Raymond

[57] **ABSTRACT**

An apparatus and method in which a tissue sample is positioned in a nuclear induction apparatus whereby selected nuclei are energized from their equilibrium states to higher energy states through nuclear magnetic resonance. By measuring the spin-lattice relaxation time and the spin-spin relaxation time as the energized nuclei return to their equilibrium states, and then comparing these relaxation times with their respective values for known normal and malignant tissue, an indication of the presence and degree of malignancy of cancerous tissue can be obtained.

16 Claims, 3 Drawing Figures

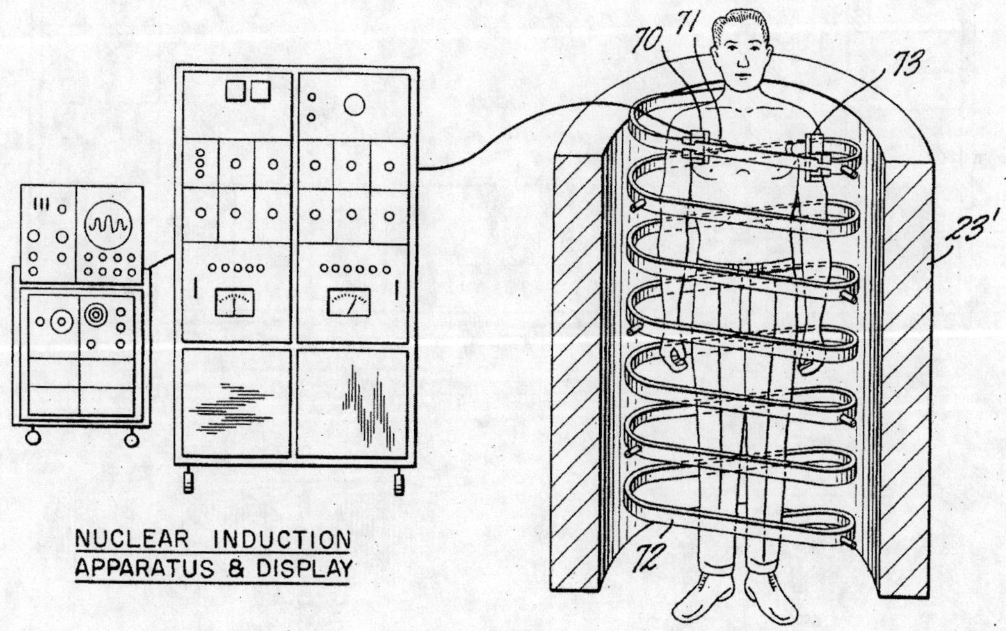

FIG. 2

WOWidea! Magnetic resonance images could uncover cancer like magic! X-rays had long been established as a diagnostic tool but were ineffective at detecting early stages of soft tissue cancers. Dr. Raymond Damadian used magnetic energy to excite the diseased cells, producing detailed pictures that save an untold numbers of lives.

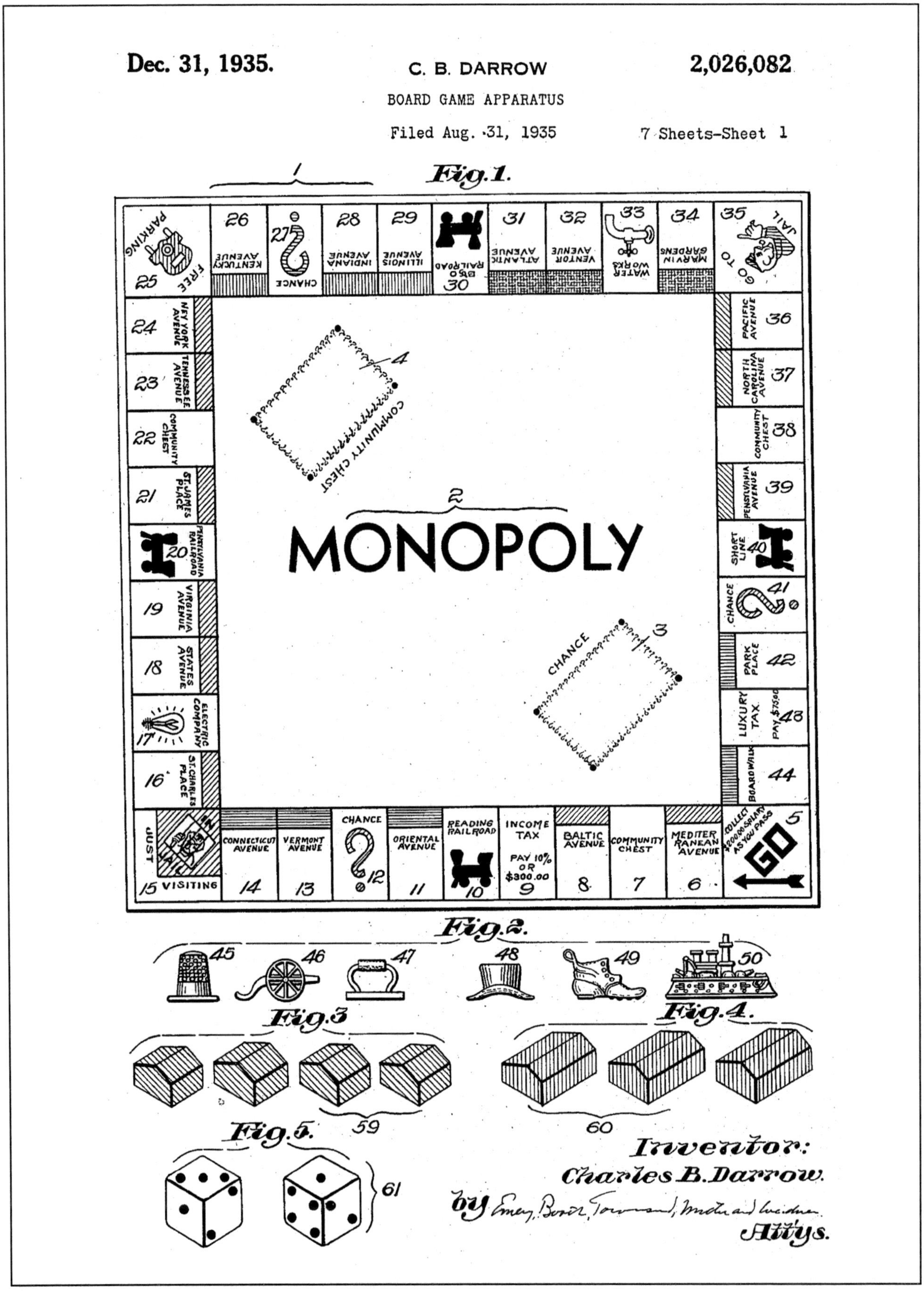

Dec. 31, 1935. C. B. DARROW 2,026,082

BOARD GAME APPARATUS

Filed Aug. 31, 1935 7 Sheets-Sheet 1

Inventor:
Charles B. Darrow.
by Emery, Booth, Townsend, Miller and Weidner
Attys.

WOWidea! Idleness can lead to success! Unemployed Philadelphia-based Charles Darrow drew upon his earlier trips to Atlantic City to create the board game as an antidote to boredom. During the height of the Depression, Parker Brothers at first turned him down saying the game was too long, too complicated, and had no clear goal.

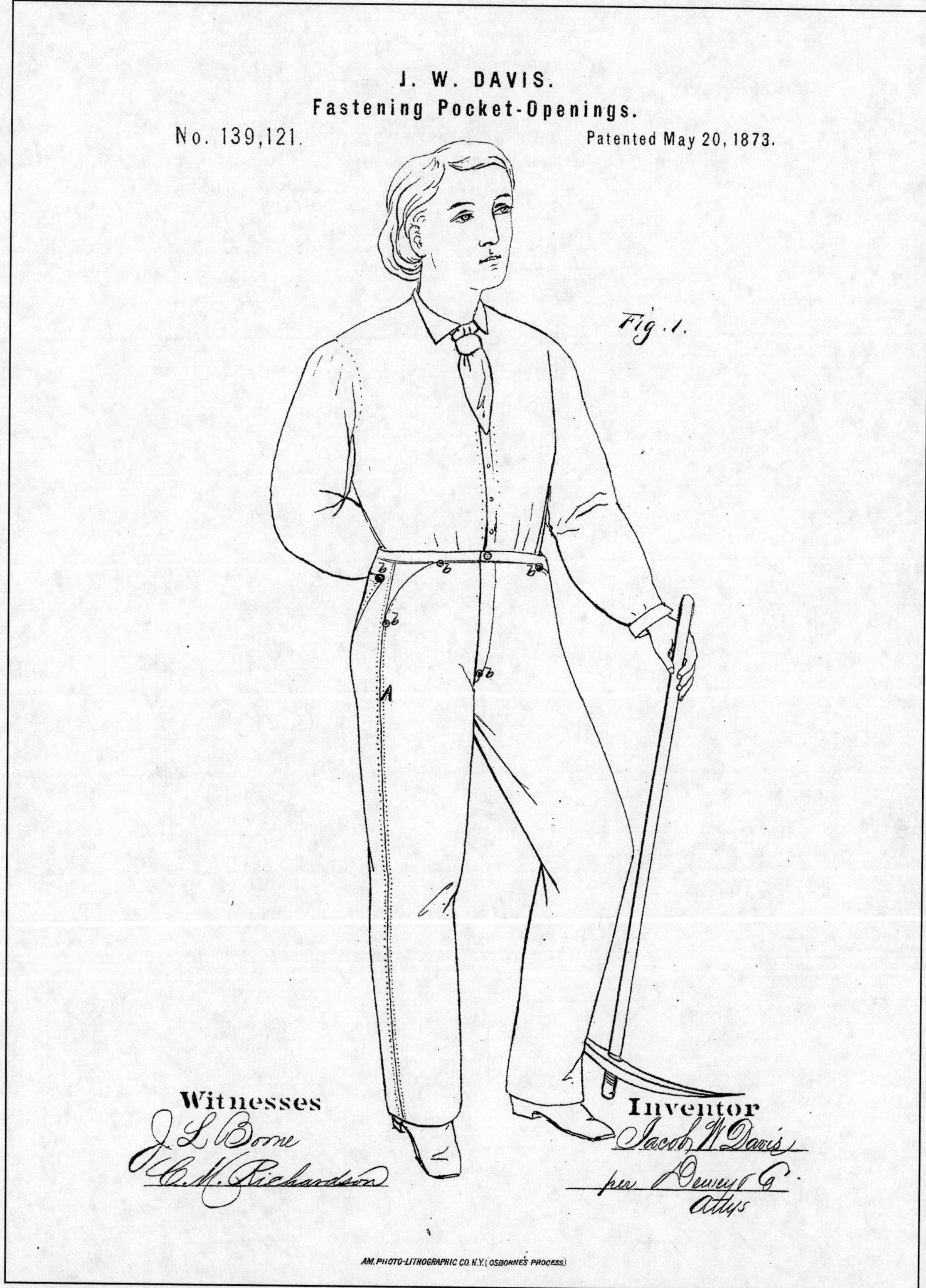

WOWidea! Miners pants that won't rip at the seams, how riveting! Jacob Davis, a Nevada tailor at the time of California's gold rush, sought the help of Levi Strauss to finance the patenting of these pants which were riveted at stressful junctures. Even the cloth was special, denim, so named for the French city of its origin (serge de Nimes).

No. 836,070. PATENTED NOV. 13, 1906.

L. DE FOREST.

OSCILLATION RESPONSIVE DEVICE.

APPLICATION FILED MAY 19, 1906.

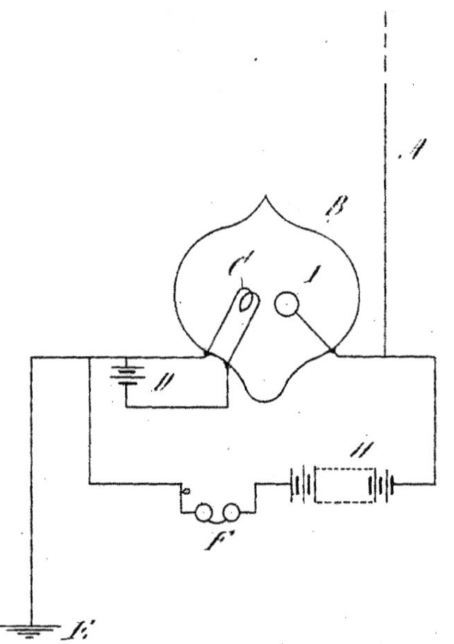

WITNESSES:

John Buckler,

Frank G. Parker

INVENTOR:

Lee de Forest

by Geo. K. Woodworth

Atty.

WOWidea! Edison didn't get it, but Lee de Forest did! An 1892 Mount Herman Boys School graduate and former bugler in the Spanish-American War, he found a way to capture mysterious electrical waves within a glass bulb void of any air. He then improved this "vacuum tube", amplifying the signals, ensuring broadcast radio's success.

Sept. 13, 1955 G. DE MESTRAL 2,717,437

VELVET TYPE FABRIC AND METHOD OF PRODUCING SAME

Filed Oct. 15, 1952

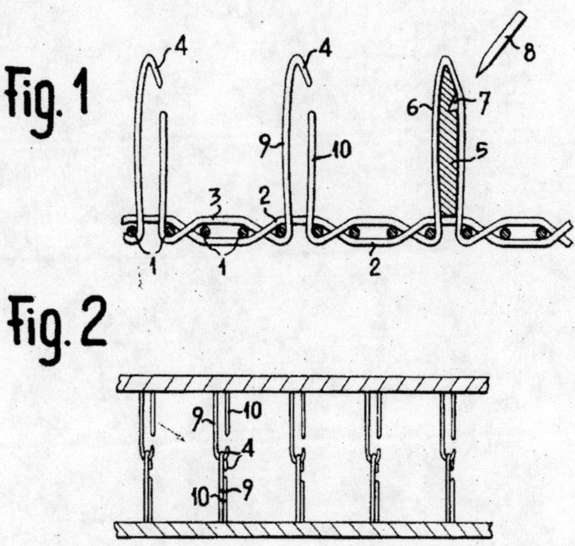

INVENTOR

George de Mestral.

BY

ATTORNEY

WOWidea! Cockleburs caught the canine's coat! George de Mestral, a Swiss engineer, was frustrated by zippers and used a walk in the park with his dog to start an eight year journey to refine his ideas. Using the catchy concept employed by nature to attach two items together, Velcro® now seems an indispensable part of many projects.

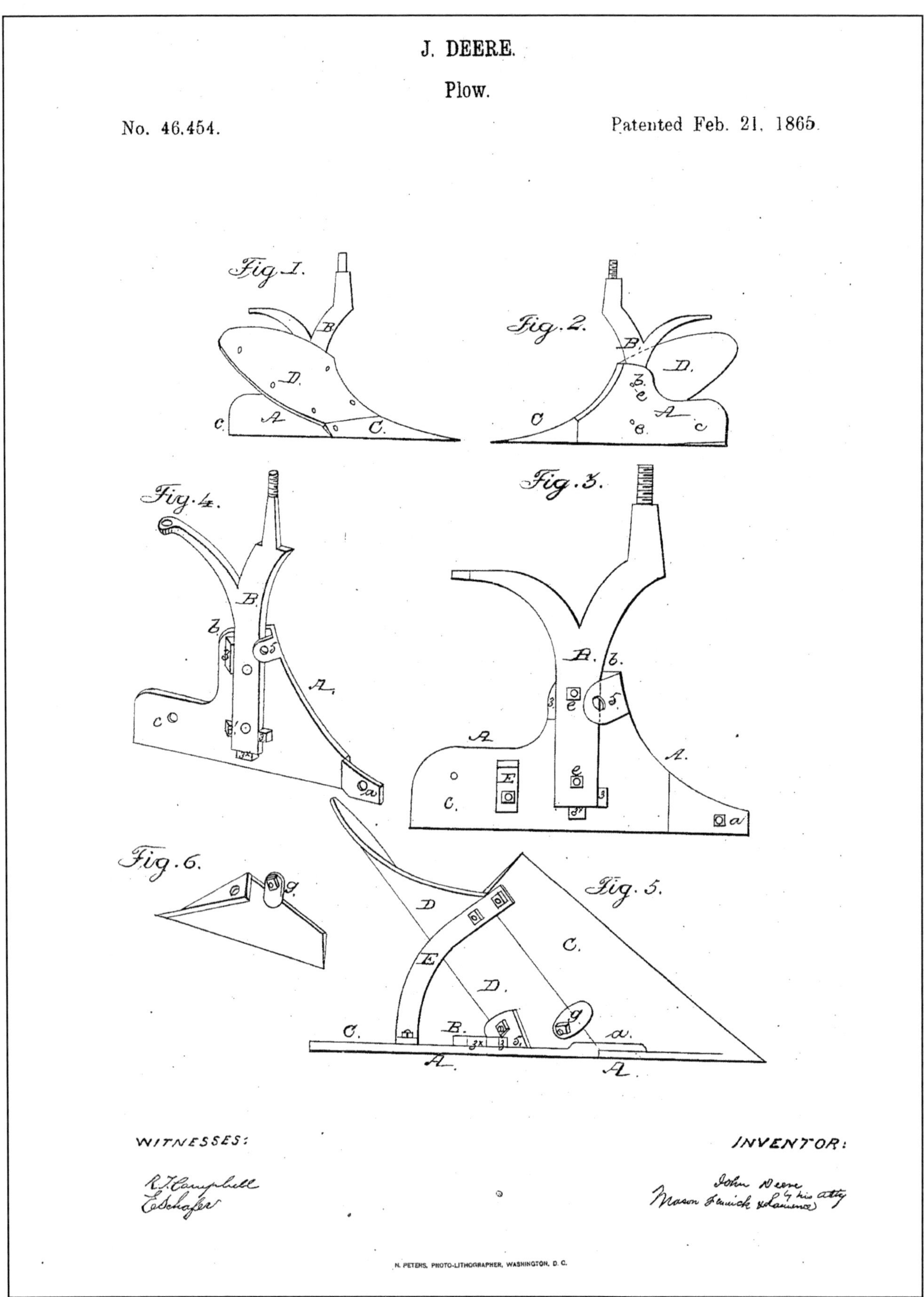

WOWidea! Using old saw blades to cut the earth was a sharp invention! The plow is an ancient concept dating back to the creation of the wheel, lever, and screw, yet with some ingenuity it was transformed. By simply covering a plow's moldboard with "recycled" steel, John Deere became a household name among farmers worldwide.

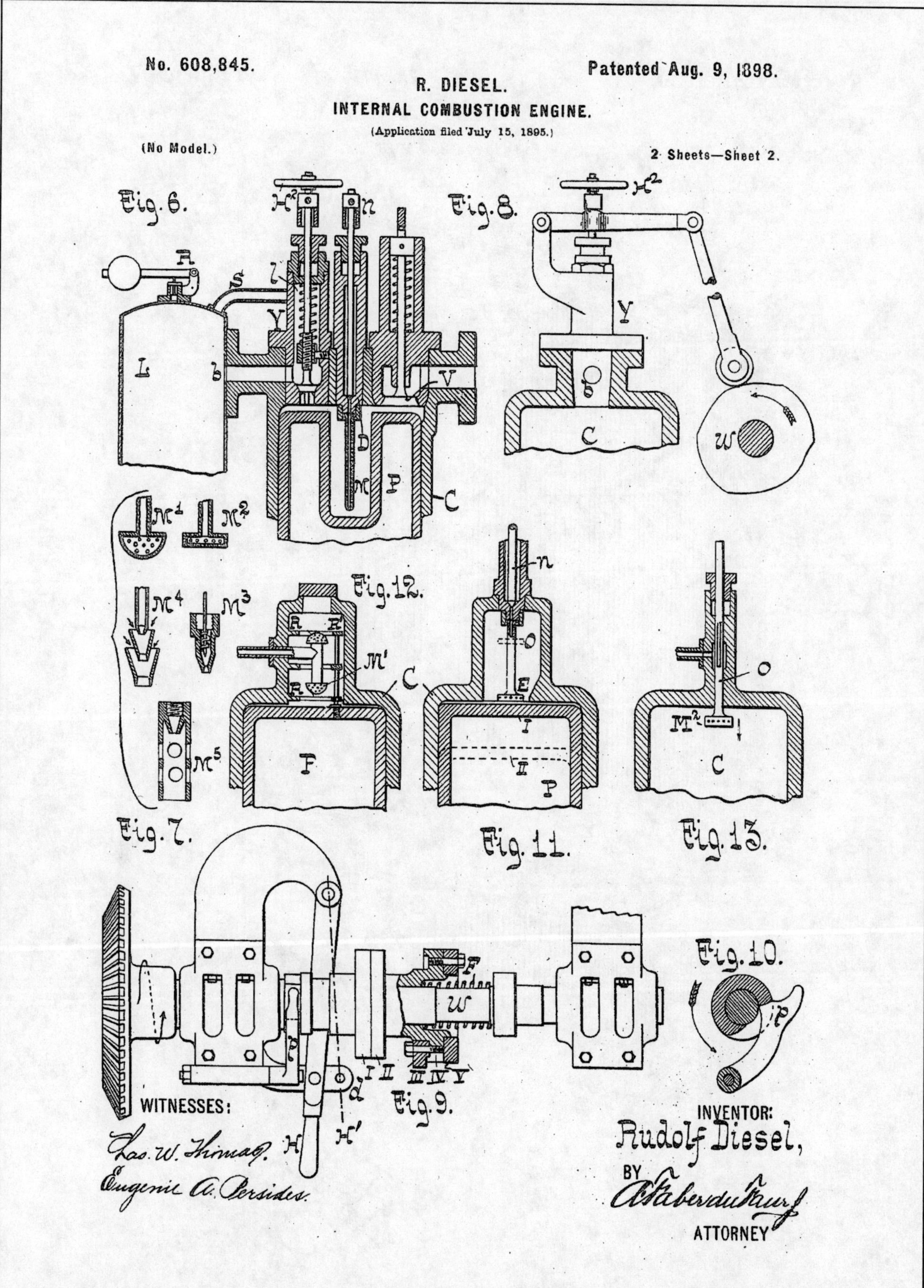

WOWidea! No spark plug, starter or carburetor made this engine cheaper and more reliable! The already-invented 2 and 4-cycle gasoline engines were taken to task by Rudolph Diesel's more efficient design. Brewery mogul Adolphus Busch bought the rights to this patent. Yet later, in financial distress, Diesel mysteriously vanished.

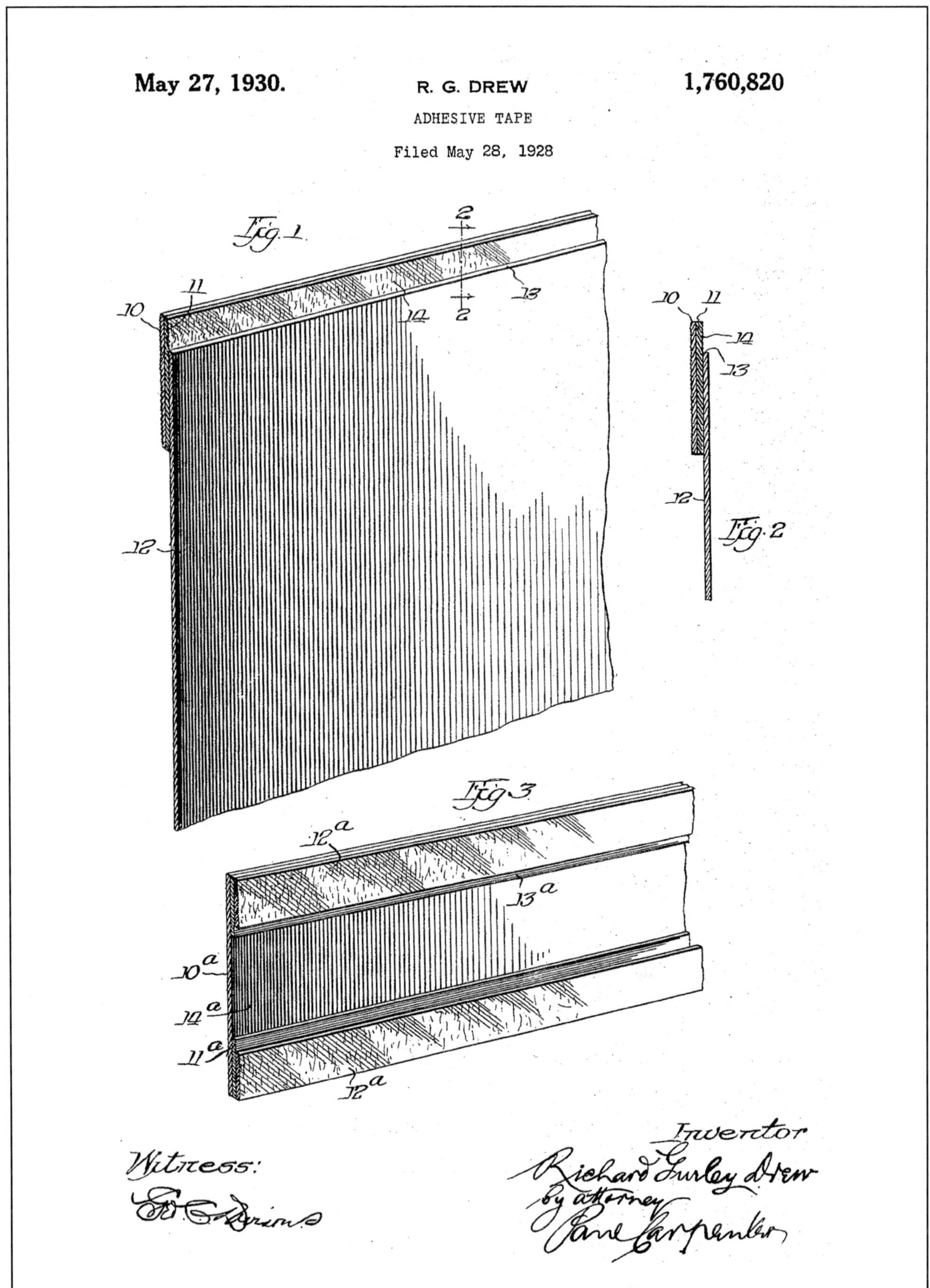

WOWidea! Ignore the boss to succeed! Richard Drew worked for 3M® in developing better adhesive masking tape and sandpapers for the automobile industry. Disobeying his manager, he experimented with a relatively new product called Cellophane®. Combining the clear film with the right kind of adhesive created Scotch® Tape.

April 27, 1965 R. C. DUNCAN ETAL 3,180,335

DISPOSABLE DIAPER

Filed July 17, 1961 2 Sheets-Sheet 1

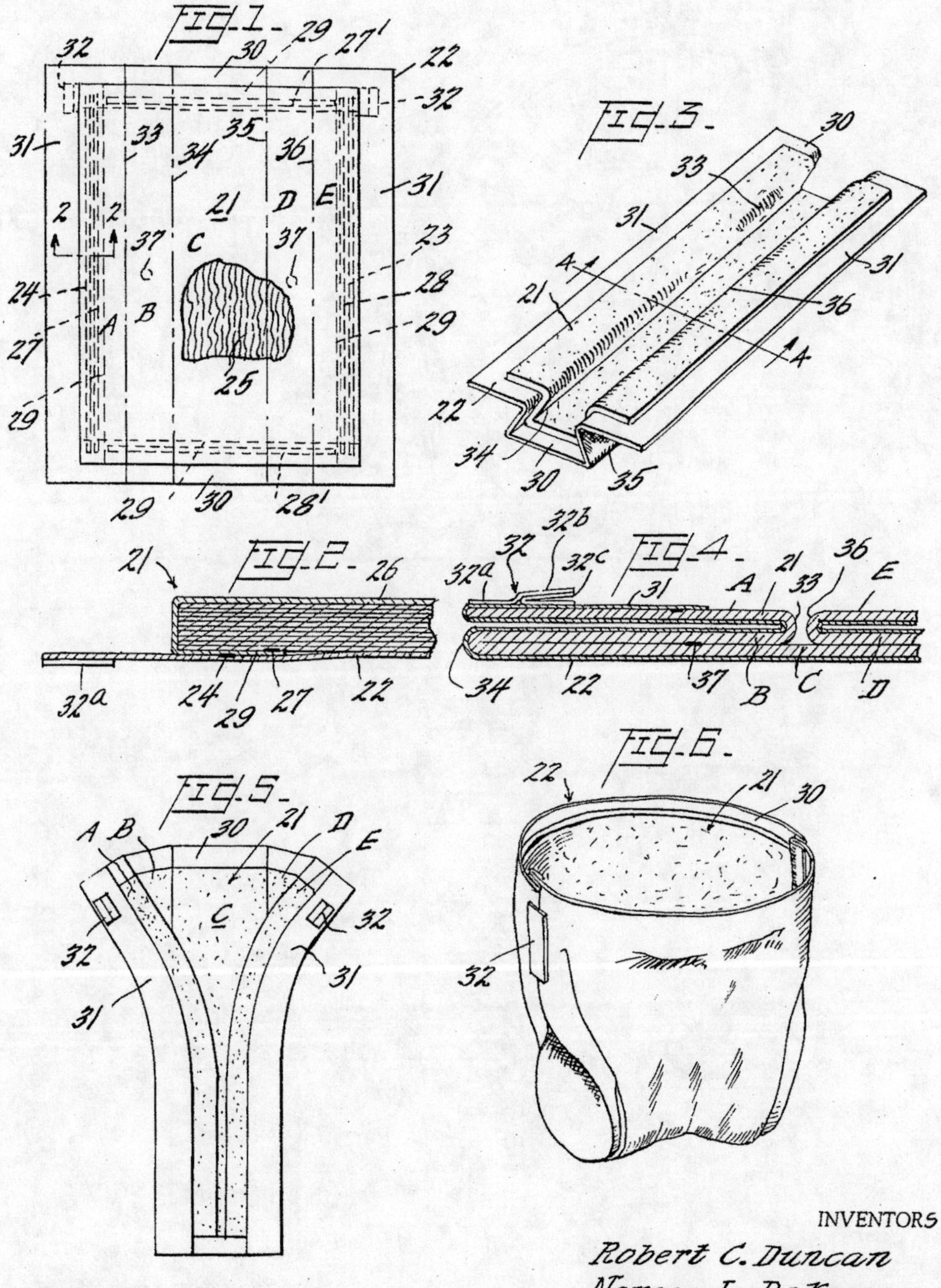

INVENTORS
Robert C. Duncan
Norma L. Baker,

BY John V. Gorman
ATTORNEY

WOWidea! Most times it takes a man and a woman working together to solves life's little problems! So it was for Robert Duncan and Norma Baker, who as employees of Proctor and Gamble, created the first disposable diaper to become commercially successful. Who was happier, the child that wore Pampers® or the parent that held them?

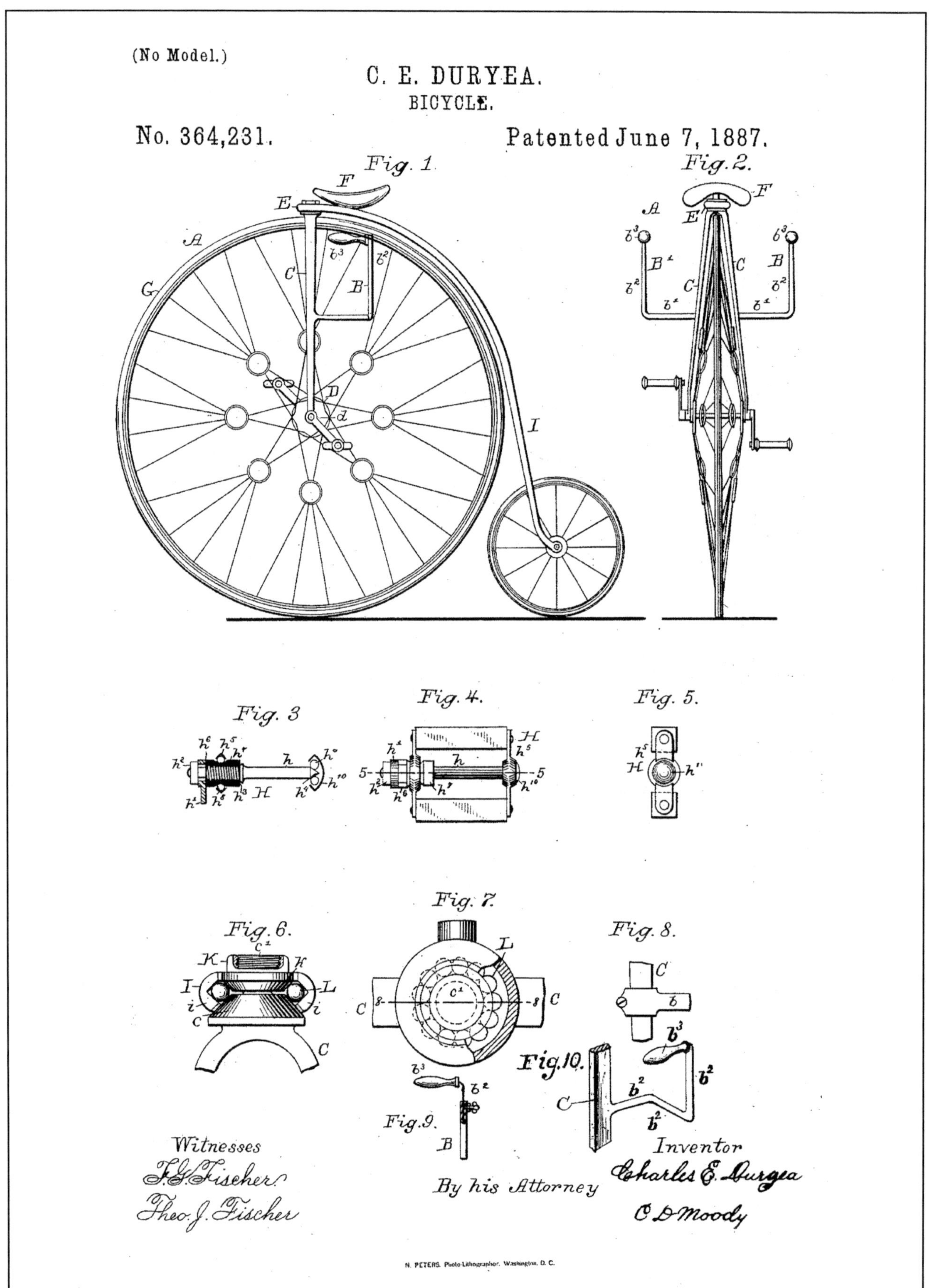

WOWidea! This bicycle led to America's first production automobile! Charles Duryea, along with his younger brother, Frank, were originally bike assemblers but were soon inspired to add a gasoline engine they saw at the 1886 Ohio State Fair. In 1895 Duryea won North America's first road race against Karl Benz in Chicago, IL.

(No Model.) 3 Sheets—Sheet 1.

G. EASTMAN.

CAMERA.

No. 388,850. Patented Sept. 4, 1888.

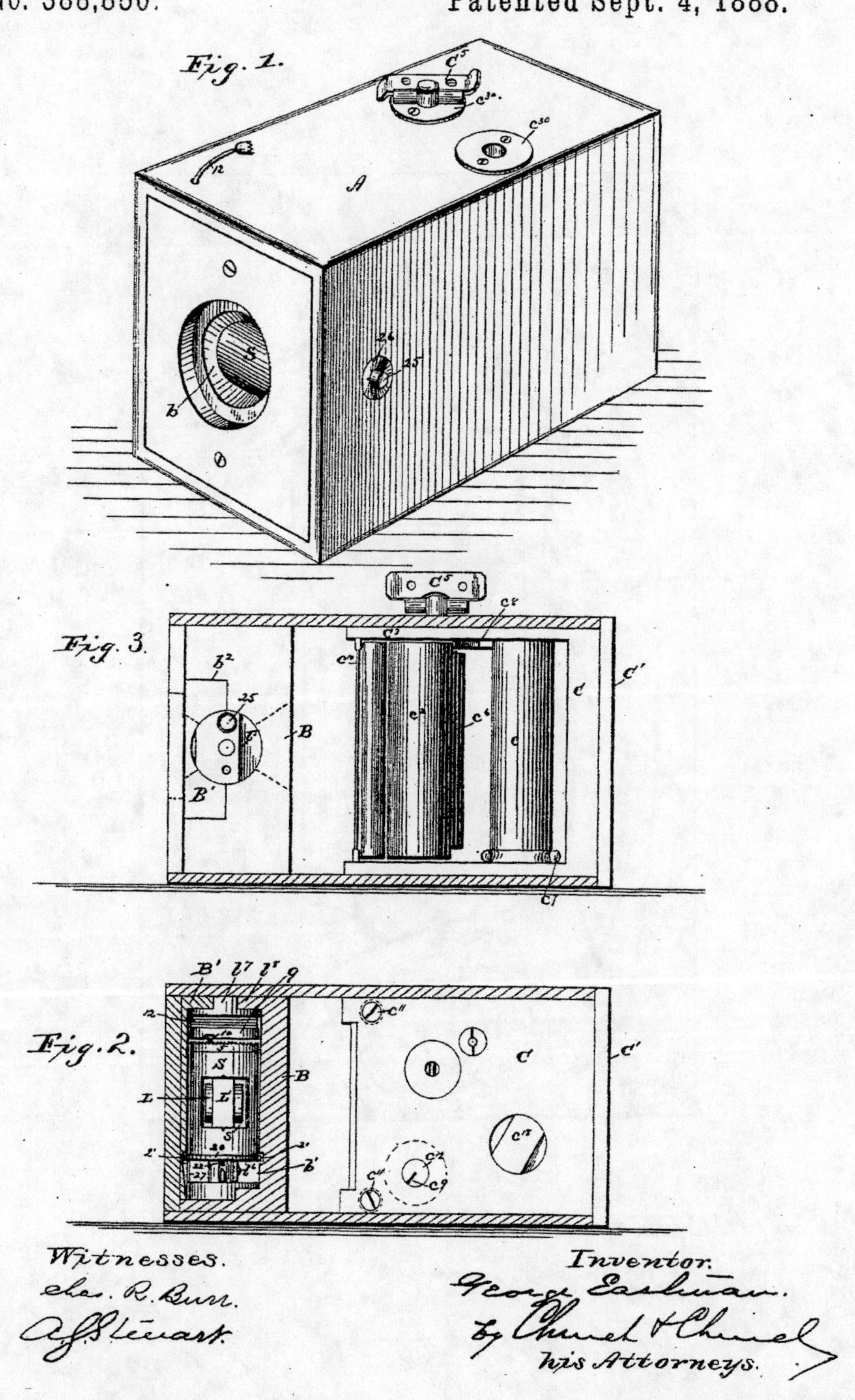

WOWidea! You pressed the button- they did the rest! George Eastman, a one-time bank bookkeeper, discovered a formula for success: make it fun and easy! He sold his Kodak® camera loaded with film and, when the pictures had been taken, the buyer returned it to the company. They sent back all the prints and loaded it with a new roll of film.

T. A. EDISON.

Electric Vote-Recorder.

No. 90,646. Patented June 1, 1869.

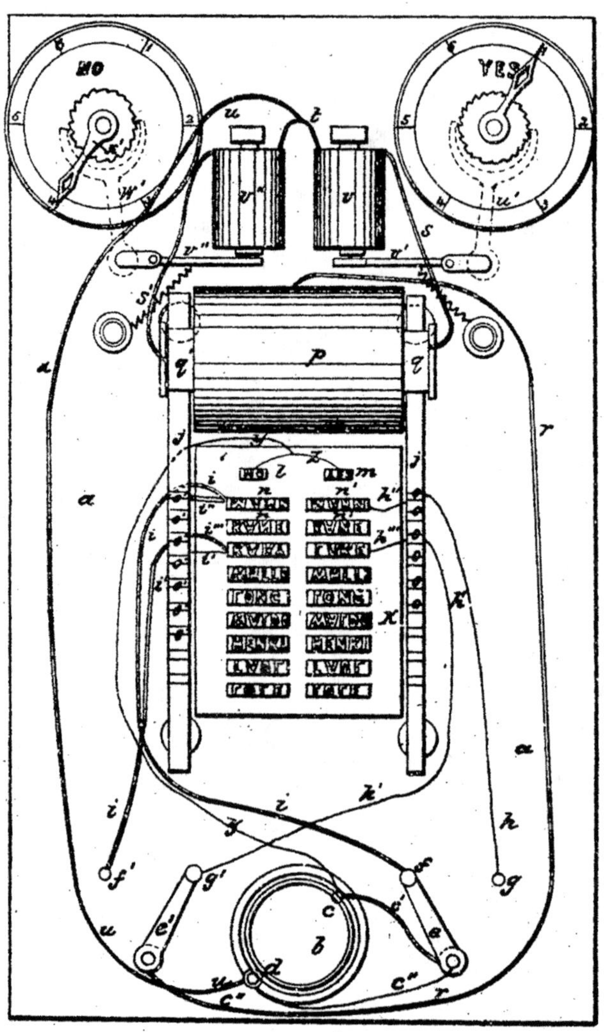

Witnesses.

Carroll D. Wright

DeWitt C. Roberts

Inventor.

Thomas A Edison.

WOWidea! Learn how your Senator voted! It was a recipe for disaster. Thomas Alva Edison envisioned the first of his 1,093 United States patents as a creative way to register and keep record of the votes made on particular bills brought before Congress. For "curious" reasons, politicians felt very differently and it was destined to be a flop.

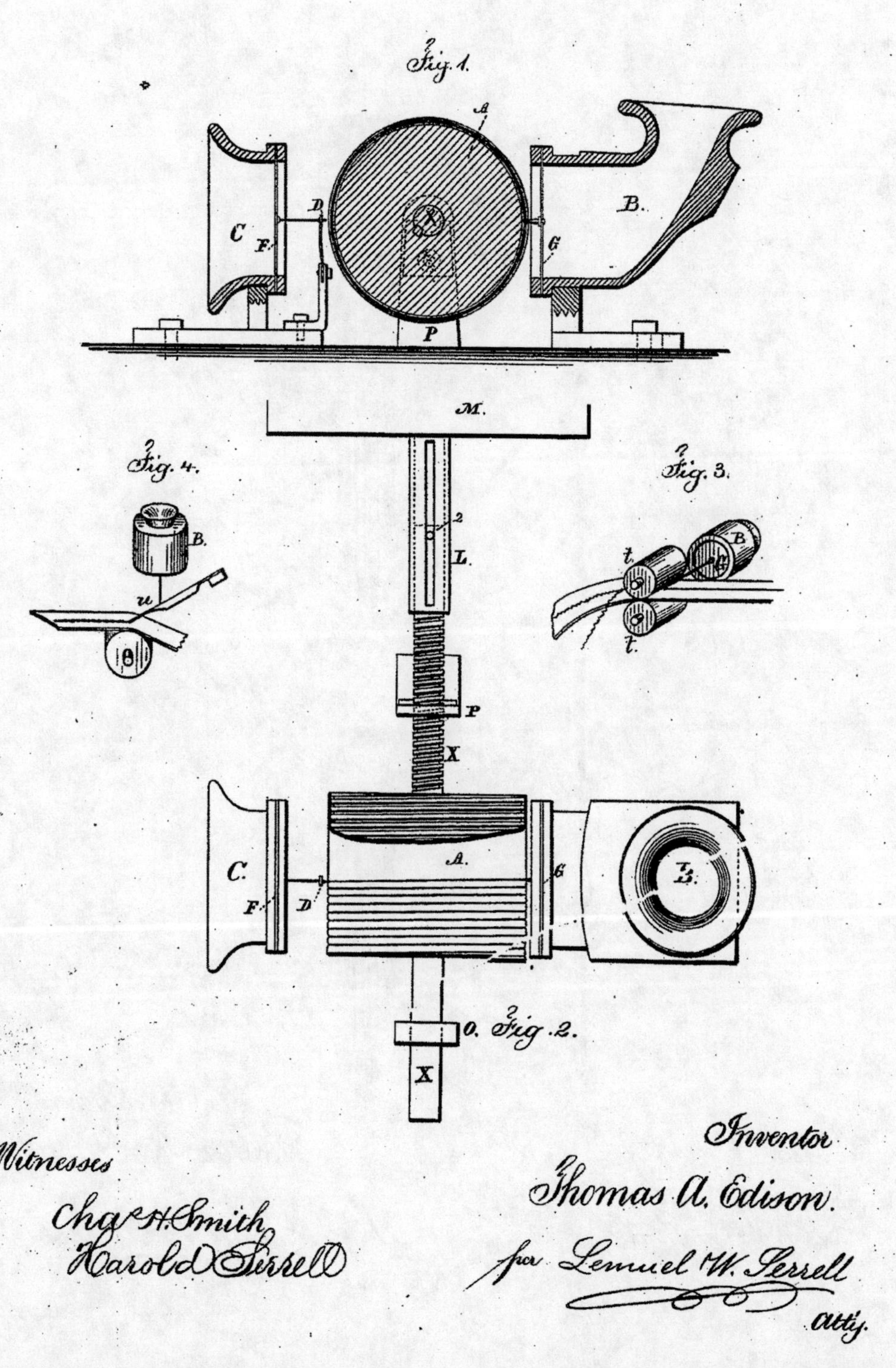

WOWidea! Good vibrations! While perfecting a telegraph transmission system, Thomas Edison devised a rotating tinfoil sheet which he indented with sound wave vibrations which could be played back. His first words? "Mary had a little lamb". This lead to his development of grooved wax cylinders, giving birth to the recording industry.

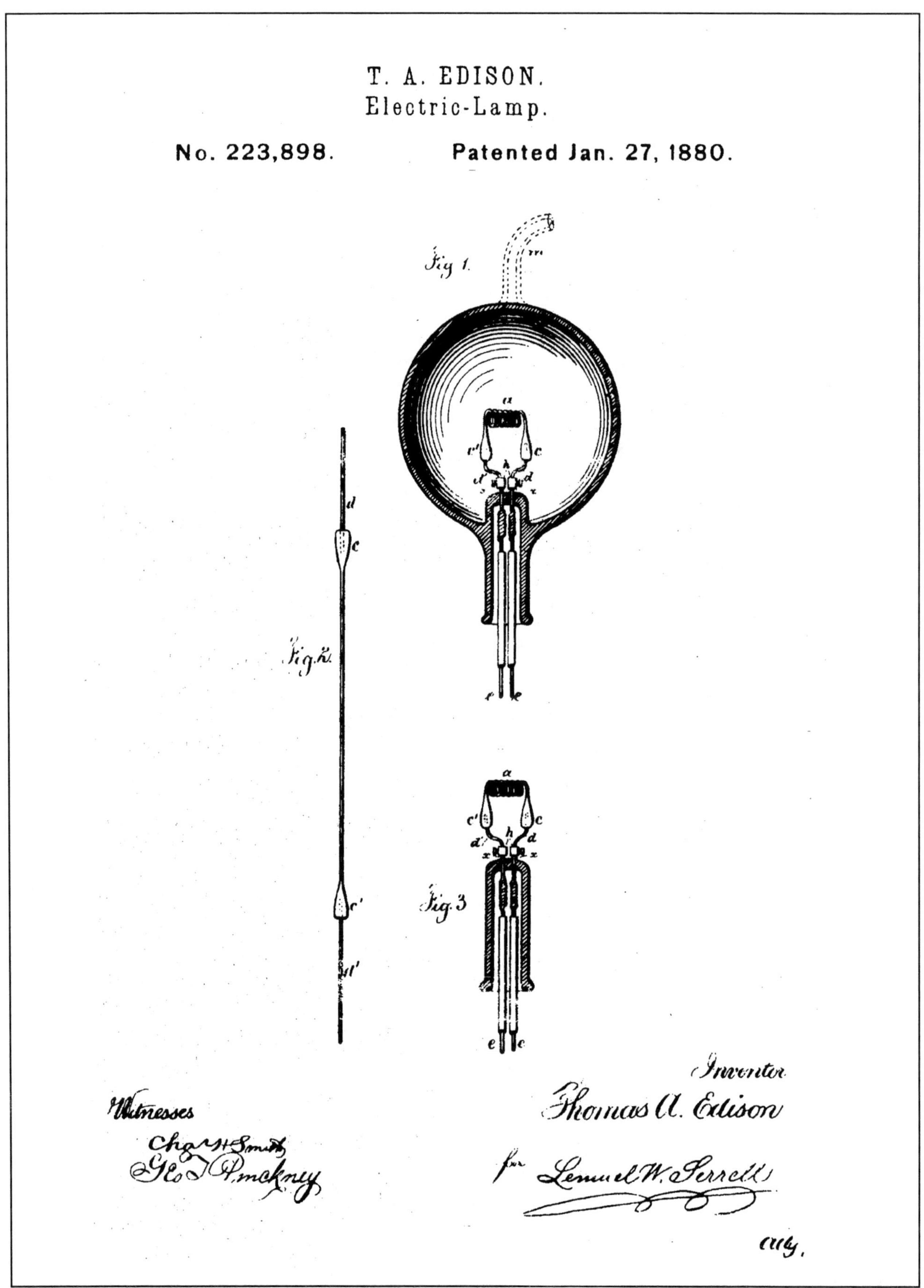

WOWidea! 1% inspiration, 99% perspiration! Thomas Edison searched the world over and experimented for years before settling on something simple, sewing thread, which he "carbonized" to make a long-lasting lamp filament. He shared the spotlight with Englishman Joseph Swan who had created a similar light bulb 20 years earlier.

(No Model.) 4 Sheets—Sheet 1.

T. A. EDISON.

APPARATUS FOR EXHIBITING PHOTOGRAPHS OF MOVING OBJECTS.

No. 493,426. Patented Mar. 14, 1893.

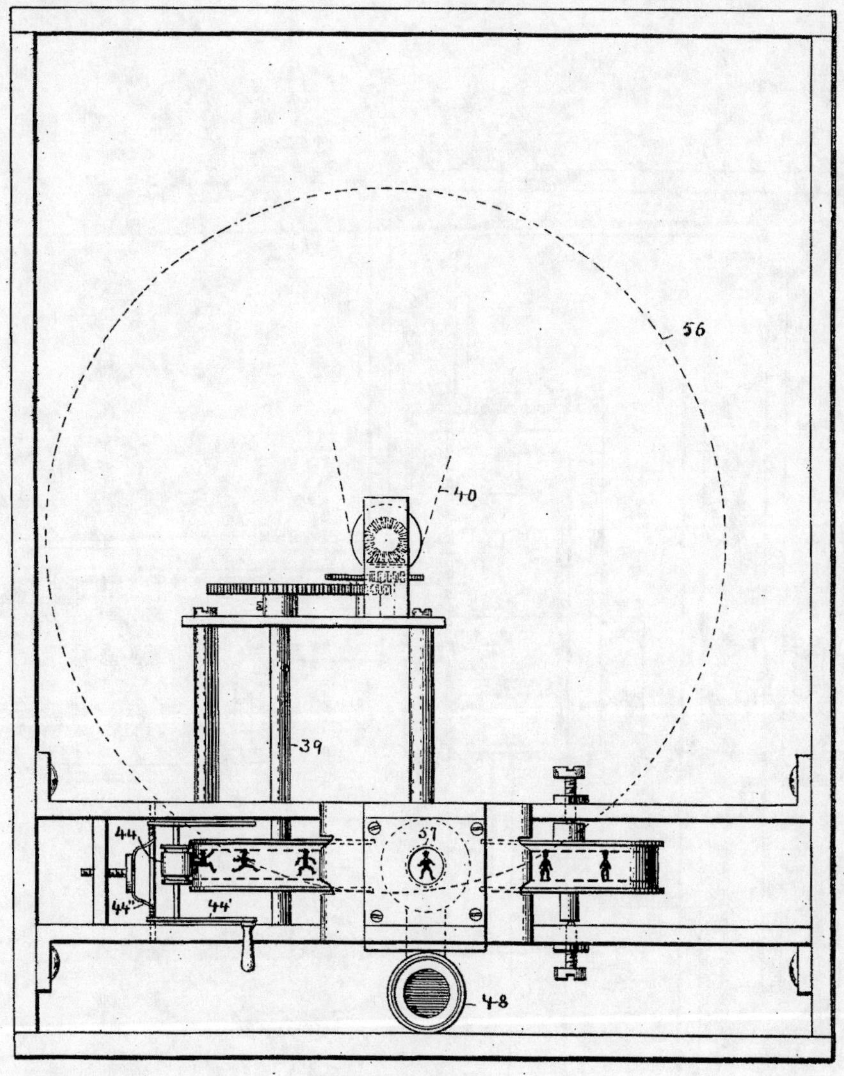

Witnesses
Norris G. Clark.
N. F. Oberlé

Inventor
T. A. Edison,
By his Attorneys
Dyer & Seely

WOWidea! Project a series of illuminated still pictures to simulate motion! Thomas Edison used the phenomenon "persistence of vision" to create both a movie camera and projector. Building on Eastman Kodak's work in celluloid, he created the perforated 35 millimeter standard-width film and the world's first motion picture studio.

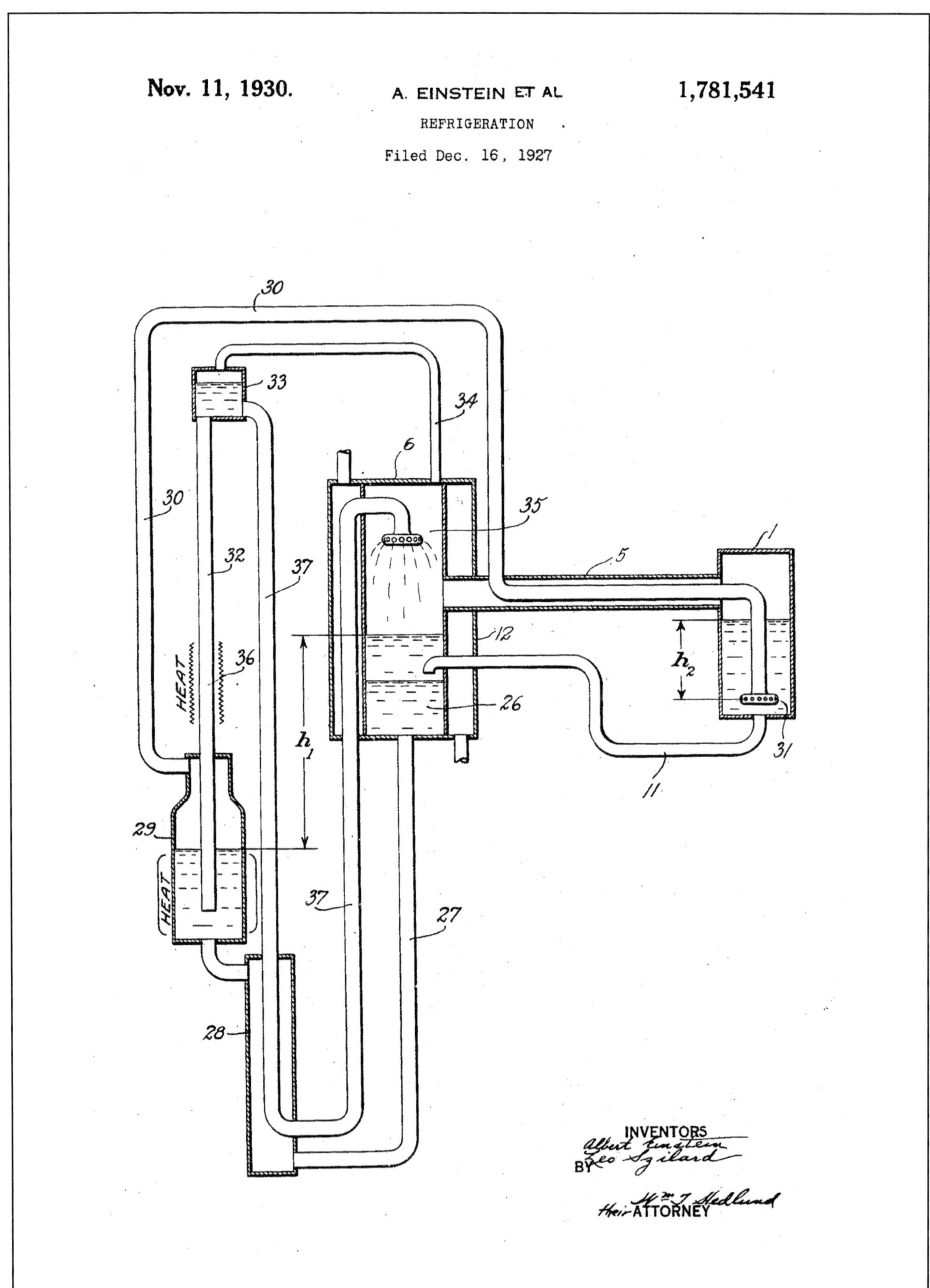

WOWidea! Relatively speaking, this idea was pretty cool! Albert Einstein, a patent examiner before leaving his native Europe, is considered by many, the leading thinker of the 20th century. He patented 45 such refrigerant contraptions. Leo Szilard, his co-inventor, went on to patent the basic concept of a nuclear chain reaction.

United States Patent Office

3,625,711
Patented Dec. 7, 1971

1

3,625,711
CYCLAMATE-FREE ARTIFICIAL SWEETENER
Marvin E. Eisenstadt, Belle Harbor, N.Y., assignor to Cumberland Packing Corporation, Brooklyn, N.Y.
No Drawing. Continuation-in-part of application Ser. No. 822,024, May 5, 1969, which is a continuation-in-part of application Ser. No. 794,767, Jan. 28, 1969, both now abandoned. This application Nov. 5, 1969, Ser. No. 874,413
Int. Cl. A231 *1/26*
U.S. Cl. 99—141 A **8 Claims**

ABSTRACT OF THE DISCLOSURE

This invention relates to a saccharine artificial sweetening composition, and more particularly to a sweetening composition, which is free of cyclamate and contains only saccharine as the artificial sweetener but which contains additives which eliminate the undesired bitter aftertaste of saccharine. The additives which are used according to this invention, and which must be used in combination because the only sweetening agent is saccharine, are lactose or dextrose, but preferably lactose, and cream of tartar powder.

This application is a continuation-in-part of Ser. No. 822,024, filed May 5, 1969, now abandoned, which was a continuation-in-part of Ser. No. 794,767, filed Jan. 28, 1969, now abandoned.

BACKGROUND OF THE INVENTION

The use of artificial sweeteners in place of sugar for reduction of caloric intake, for medical reasons, and the like is well known. The most common artificial sweeteners are the saccharines and the cyclamates (which latter are sold under the trademark "Sucaryl").

Both of the above groups of substances are much sweeter than sugar and have no caloric value. However, both of the groups of substances suffer from the disadvantage of leaving a bitter aftertaste in the mouth of the user, and very often a saccharine and a cyclamate are used in order to lower the degree of bitter aftertaste of the sweeteners.

The cyclamates have about 30 times the sweetening power of pure sugar (referring to sugar what is meant always is the normal cane sugar or beet sugar which is used commercially for sweetening and which actually consists mainly of sucrose).

The saccharines have a much higher degree of sweetening power, namely about 300 times the sweetening power of sugar. The saccharines, however, have an even greater bitter aftertaste than the cyclamates, and the saccharines are most often not used alone, but rather in admixture with a cyclamate.

U.S. Pat. No. 3,259,506 is directed to combinations of saccharine and cyclamate with lactose for the purpose of minimizing the bitter aftertaste of the two artificial sweeteners. However, using lactose for this purpose it is still necessary to use a combination of both saccharine and cyclamate in order to obtain a composition which does not have a bitter aftertaste but which still has the desired sweetness. The reason for this is that lactose, while it has the property, when used in sufficient amount, to depress or completely destroy the bitter aftertaste of the artificial sweetener will, when too great a quantity is used alone, have the undesired effect of changing the taste of the food or beverage to which the same is applied. Since saccharine is so extremely sweet, namely 300 times as sweet as sugar, if the saccharine were used alone with lactose, such large amount of lactose would be required in order to depress the bitter aftertaste of the

2

saccharine that when the composition is used, for example to sweeten a cup of coffee, the large amount of lactose would adversely affect the taste of the coffee. This was avoided by combining saccharine and cyclamate along with the lactose.

However, recently there have been medical reports that the cyclamates have undesired physiological side effects, and in fact the United States Food & Drug Administration has recently banned the use of cyclamates. It is therefore apparent that it is desirable to provide artificial sweetening products which do not contain cyclamates, but also which do not have the undesired bitter aftertaste of saccharine.

Generally speaking, in accordance with the present invention, a composition is provided of saccharine, dextrose or lactose, but preferably dextrose, and powdered cream of tartar. Dextrose, of course, is a natural sweetening sugar which has practically the same sweetening effect as normal sugar (by which is meant the normal cane sugar or beet sugar, and which is chemically called sucrose) and can itself be used as a sweetener. Lactose, on the other hand, although it is chemically known as milk sugar, is not a true sweetening agent because it has less than one fifth the sweetness of natural sugar (sucrose). However, in the small amounts used according to the present invention, neither the dextrose nor the lactose could act as a sweetener, the sweetening effect actually being provided by the saccharine. However, the dextrose or the lactose, when used together with the cream of tartar powder, make it possible to provide a composition which avoids the bitter aftertaste of saccharine, while using saccharine alone as the artificial sweetener. Dextrose alone cannot be used for this purpose because the dextrose alone will not sufficiently mask the bitter aftertaste of the saccharine. If lactose alone, on the other hand, were used for this purpose, the bitter aftertaste of the saccharine would be masked, however, there would be too much lactose present, and this would have an adverse effect on the taste of the foods to which the sweetening composition is applied.

While either dextrose or lactose, or even a mixture of dextrose and lactose, can be used with the cream of tartar and a saccharine artificial sweetener to provide a sweetening composition with desirable sweetness and without bitter aftertaste, best results from the standpoint of achieving most nearly true sugar taste, are obtained by the use of lactose alone plus the cream of tartar and saccharine.

It is accordingly a primary object of the present invention to provide a sweetening composition which contains only a saccharine as the artifical sweetener but which does not have the bitter aftertaste of the saccharine while being able to be used with all types of foods without having any undesired effect on the normal taste of the food.

It is yet a further object of the present invention to provide compositions of saccharine, lactose and/or dextrose, and cream of tartar powder, which have no undesired bitter aftertaste and which can be used with all types of food while providing only a sweetening effect of the food which is akin to that of natural sugar.

Other objects and advantages of the present invention will be apparent from a further reading of the specification and of the appended claims.

The term "saccharine artificial sweetener" as used throughout the specification and claims of this case is meant to refer to saccharine itself and the salts thereof such as sodium saccharine, potassium saccharine, etc. Cream of tartar is of course also known as potassium bitartrate.

In accordance with the present invention, the saccharine artificial sweetener is mixed with the cream of

WOWidea! Tea-bags were his original inspiration! Ben Eisenstadt was the first person to successfully package sugar in packets. Soon he was packaging everything including kids toys like instant Sea Monkeys®. His son Marvin joined the business in 1957 to help create one of the world's best know sugar substitutes, Sweet N'Low®.

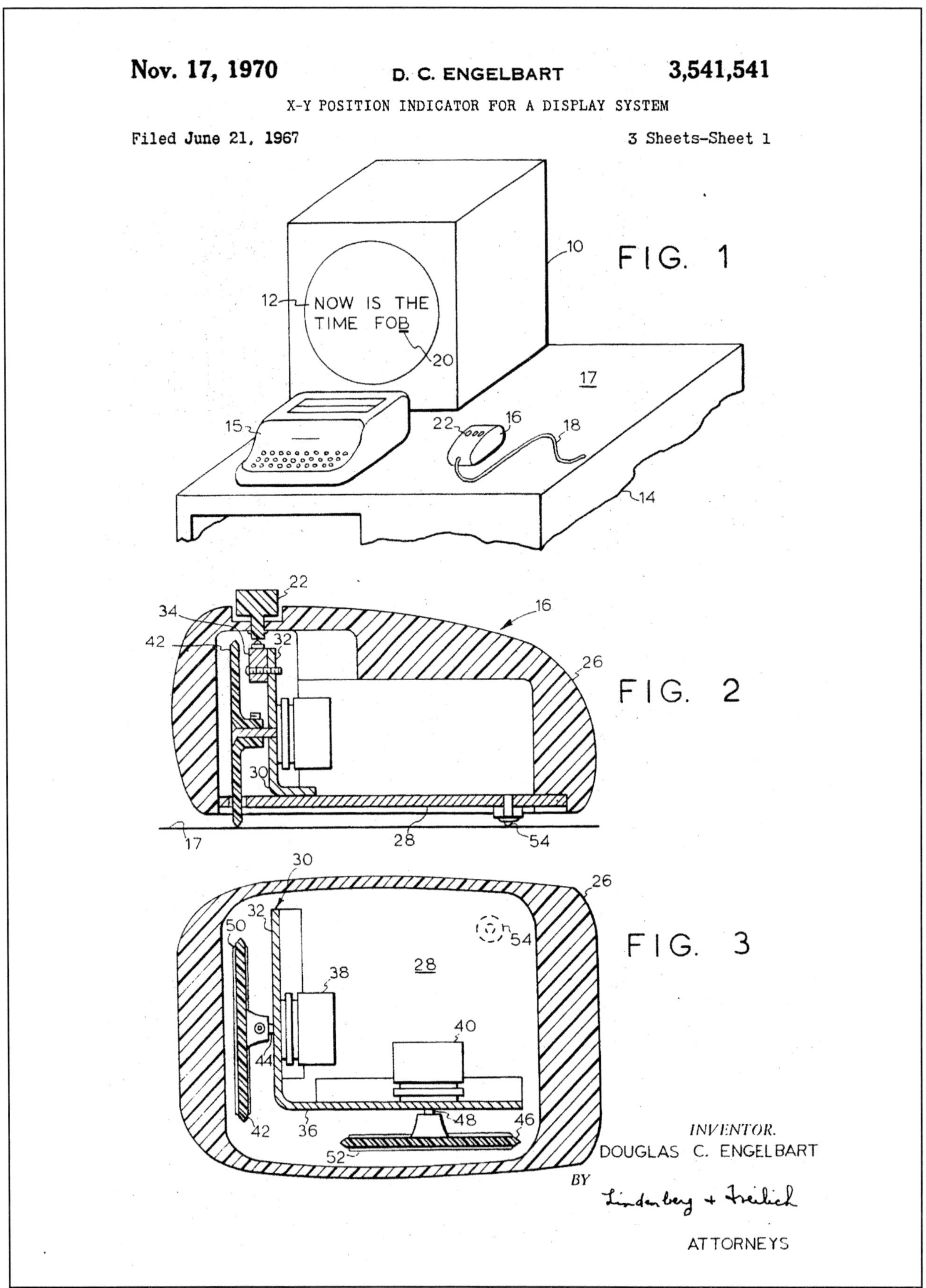

WOWidea! Imagine flying through information on a computer screen! Douglas Englebart, did just that. Using X-Y coordinates and a hand-maneuvered apparatus, he created what affectionately become known for its mouse-like looks. Steve Jobs of Apple Computer paid for the rights to be the first to integrate it into a computer, the Lisa®.

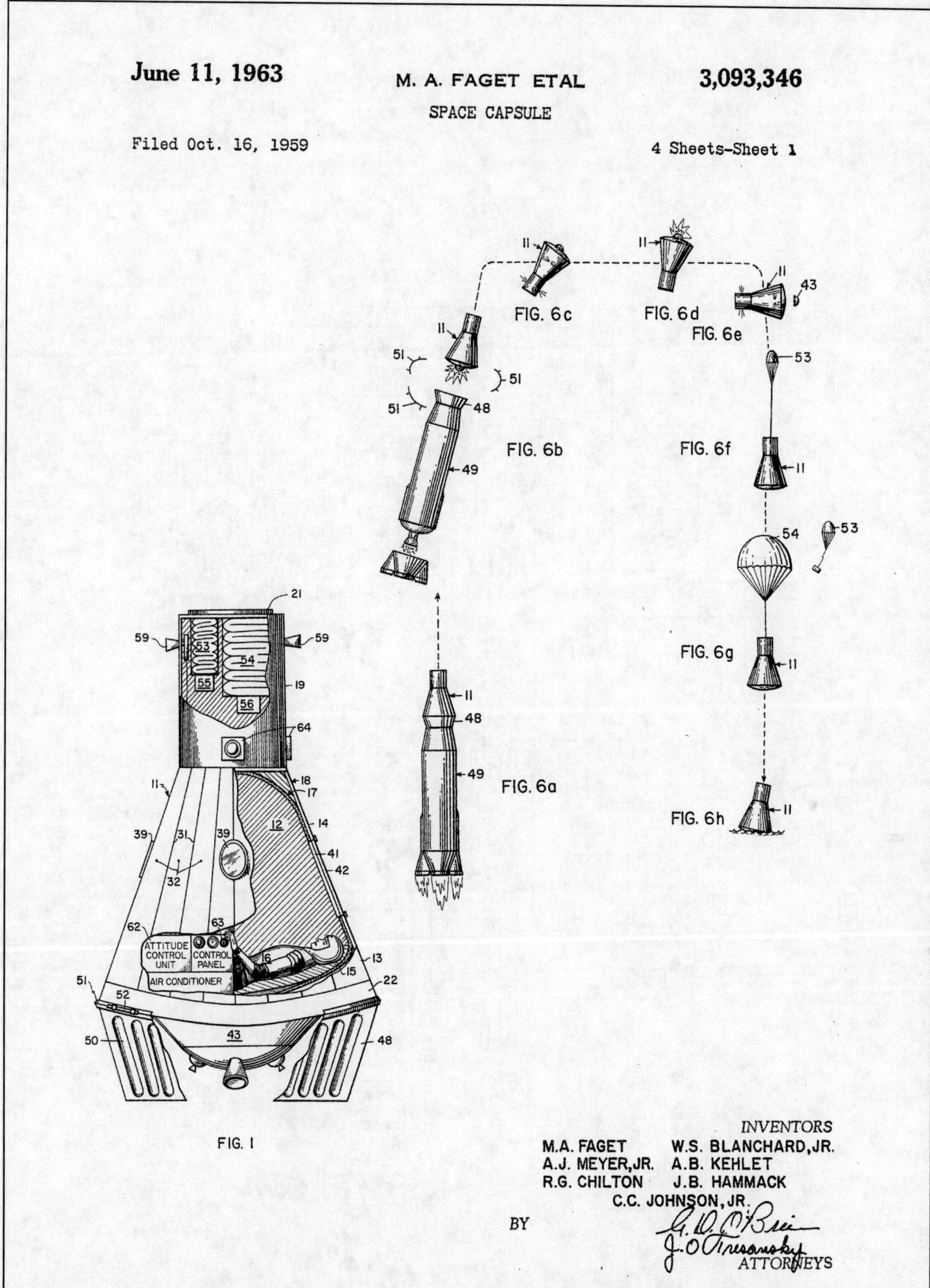

WOWidea! Put a man in space and bring him back safely! Maxime Faget, an innovator in rocketry, managed Project Mercury that put America's first man in orbit. The Gemini and Apollo space capsules followed, and were variations built upon his original design. Faget later organized the team that developed the space shuttle program.

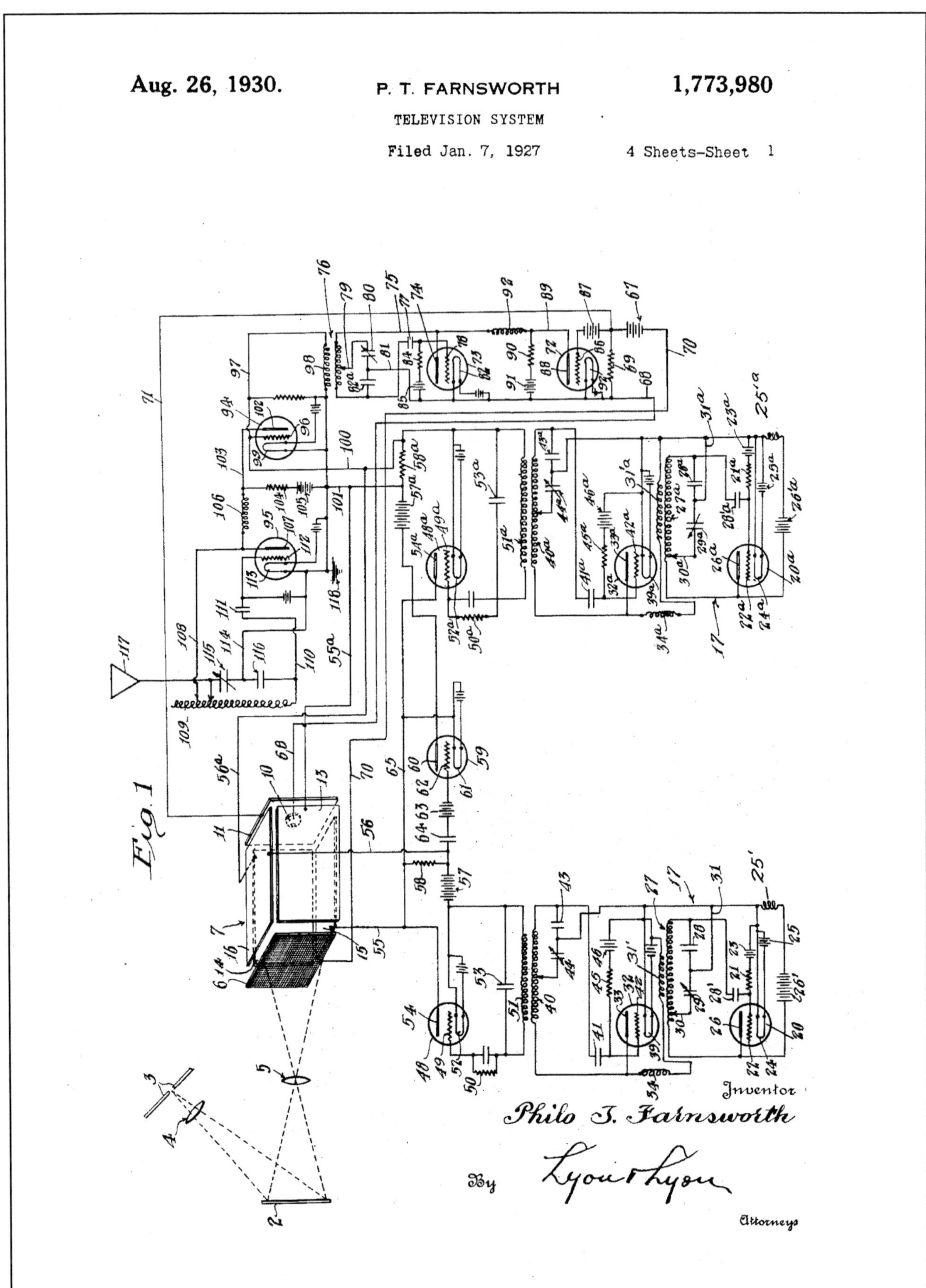

Aug. 26, 1930. P. T. FARNSWORTH 1,773,980

TELEVISION SYSTEM

Filed Jan. 7, 1927 4 Sheets-Sheet 1

WOWidea! It could all be done with electronics! Mechanical means of gathering, transmitting and displaying images had been around awhile, but at just 13, Philo Farnsworth envisioned the notion of firing electrons at a screen which would glow with an image. By the age of 21 he had formalized his theories with this landmark patent.

April 10, 1956 C. L. FENDER 2,741,146

TREMOLO DEVICE FOR STRINGED INSTRUMENTS

Filed Aug. 30, 1954

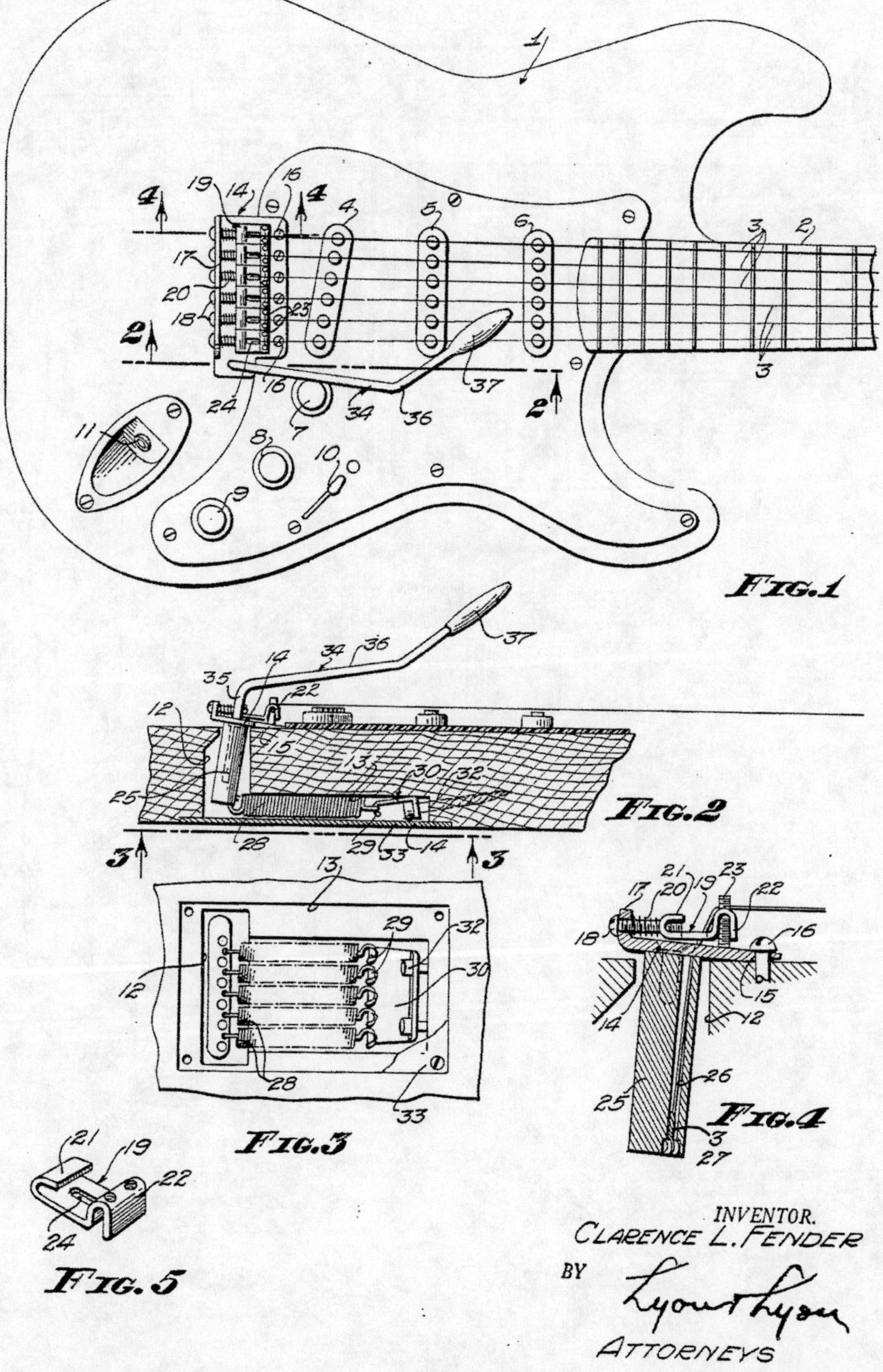

FIG. 1

FIG. 2

FIG. 3

FIG. 4

FIG. 5

INVENTOR.
CLARENCE L. FENDER

BY Lyon & Lyon
ATTORNEYS

WOWidea! Broadcaster, then Telecaster, and finally a hit, the Stratocaster®! Originally a maker of amplifiers and Hawaiian-style guitars, Clarence Leo Fender began to experiment with solid-body electric guitars. Les Paul, the guitarist and producer was also working on a similar concept at the time but Fender got his ideas to market first.

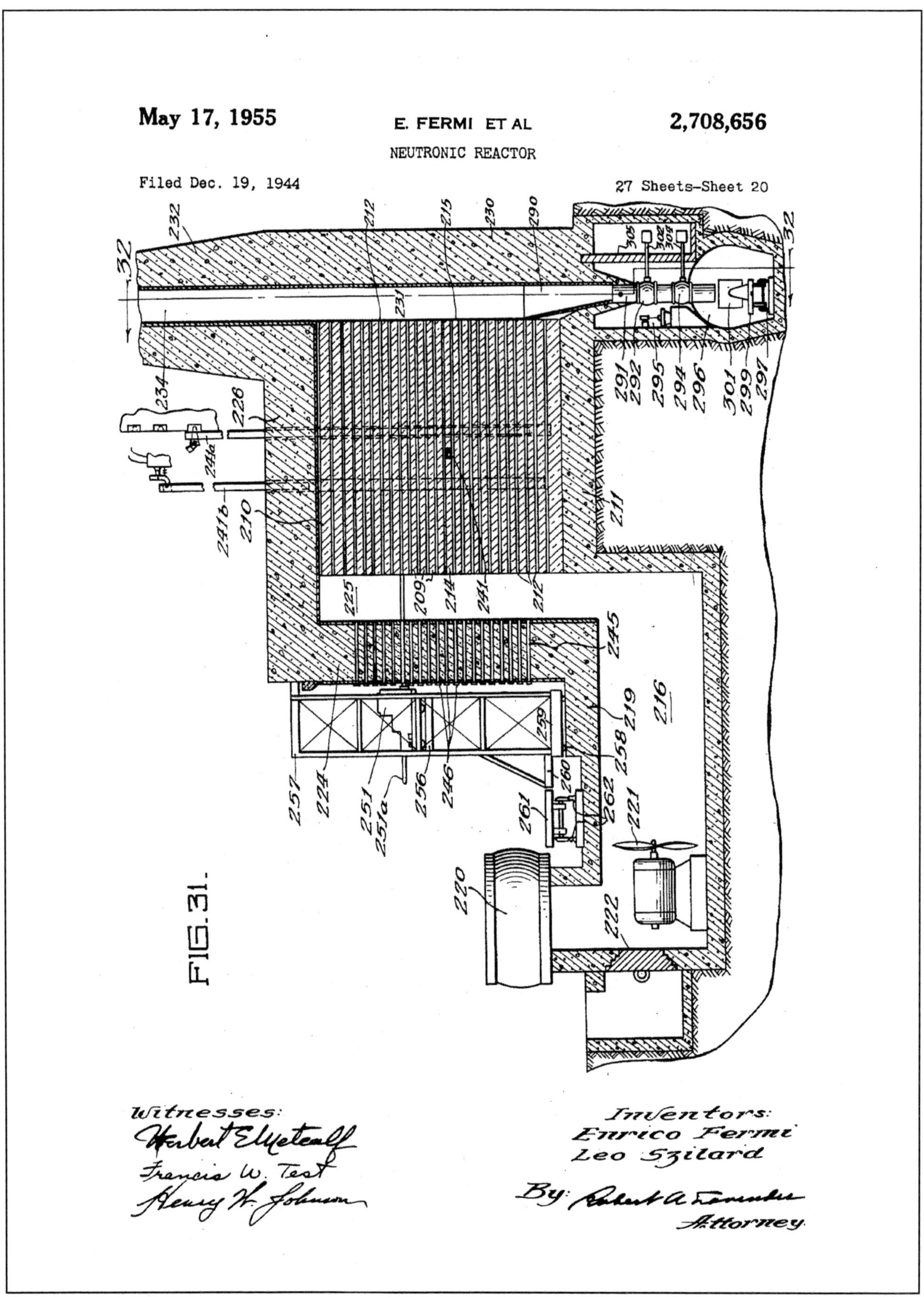

WOWidea! Harnessing the atom was the real discovery! Escaping religious and political oppression in Europe, the Nobel prize-winning Enrico Fermi found a home within the scientific community of America. He and Szilard labored to learn how to create and yet control a nuclear chain reaction. Results were as good as they could be bad.

United States Patent [19]

Fleming

[11] **Patent Number: 4,666,425**

[45] **Date of Patent: May 19, 1987**

[54] **DEVICE FOR PERFUSING AN ANIMAL HEAD**

[75] Inventor: **Chet Fleming,** St. Louis, Mo.

[73] Assignee: **The Dis Corporation,** St. Louis, Mo.

[21] Appl. No.: **809,949**

[22] Filed: **Dec. 17, 1985**

[51] **Int. Cl.**^{4} **A61M 37/00**
[52] **U.S. Cl.** .... **604/4;** 128/1 R
[58] **Field of Search** .... 604/4, 5, 6; 128/1 R

[56] **References Cited**

U.S. PATENT DOCUMENTS

4,116,589	9/1978	Rishton	604/4
4,192,302	3/1980	Boddie	604/4
4,540,399	9/1985	Litzie et al.	604/4
4,583,969	4/1986	Mortensen	604/4

OTHER PUBLICATIONS

Robt. J. White, "Brain," pp. 655–674 in Organ Preservation for Transplantation, 2nd ed., Karon & Pegg, 1981.

The Tomorrow File, by Lawrence Sanders (1975), pp. 430–432 and 459–466.

Heads, by David Osborn (1985), pp. 108–110 and 146–147.

Primary Examiner—John D. Yasko
Attorney, Agent, or Firm—Patrick Kelly

[57] **ABSTRACT**

This invention involves a device, referred to herein as a "cabinet," which provides physical and biochemical support for an animal's head which has been "discorporated" (i.e., severed from its body). This device can be used to supply a discorped head with oxygenated blood and nutrients, by means of tubes connected to arteries which pass through the neck. After circulating through the head, the deoxygenated blood returns to the cabinet by means of cannulae which are connected to veins that emerge from the neck. A series of processing components removes carbon dioxide and add oxygen to the blood. If desired, waste products and other metabolites may be removed from the blood, and nutrients, therapeutic or experimental drugs, anti-coagulants, and other substances may be added to the blood. The replenished blood is returned to the discorped head via cannulae attached to arteries. The cabinet provides physical support for the head, by means of a collar around the neck, pins attached to one or more vertebrae, or similar mechanical means.

20 Claims, 3 Drawing Figures

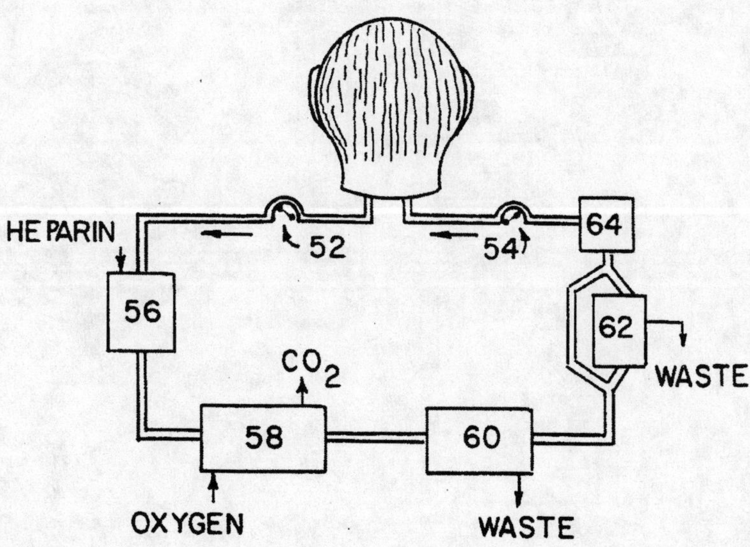

WOWidea! Science fiction might become science future! Chet Fleming, a pseudonym for a mid-west lawyer, wrote the book "If We Can Keep a Severed Head Alive..." to make a point and he succeeded. Along with cryogenics and other means of seeking immortality, Fleming used the patent process to provoke a heady debate.

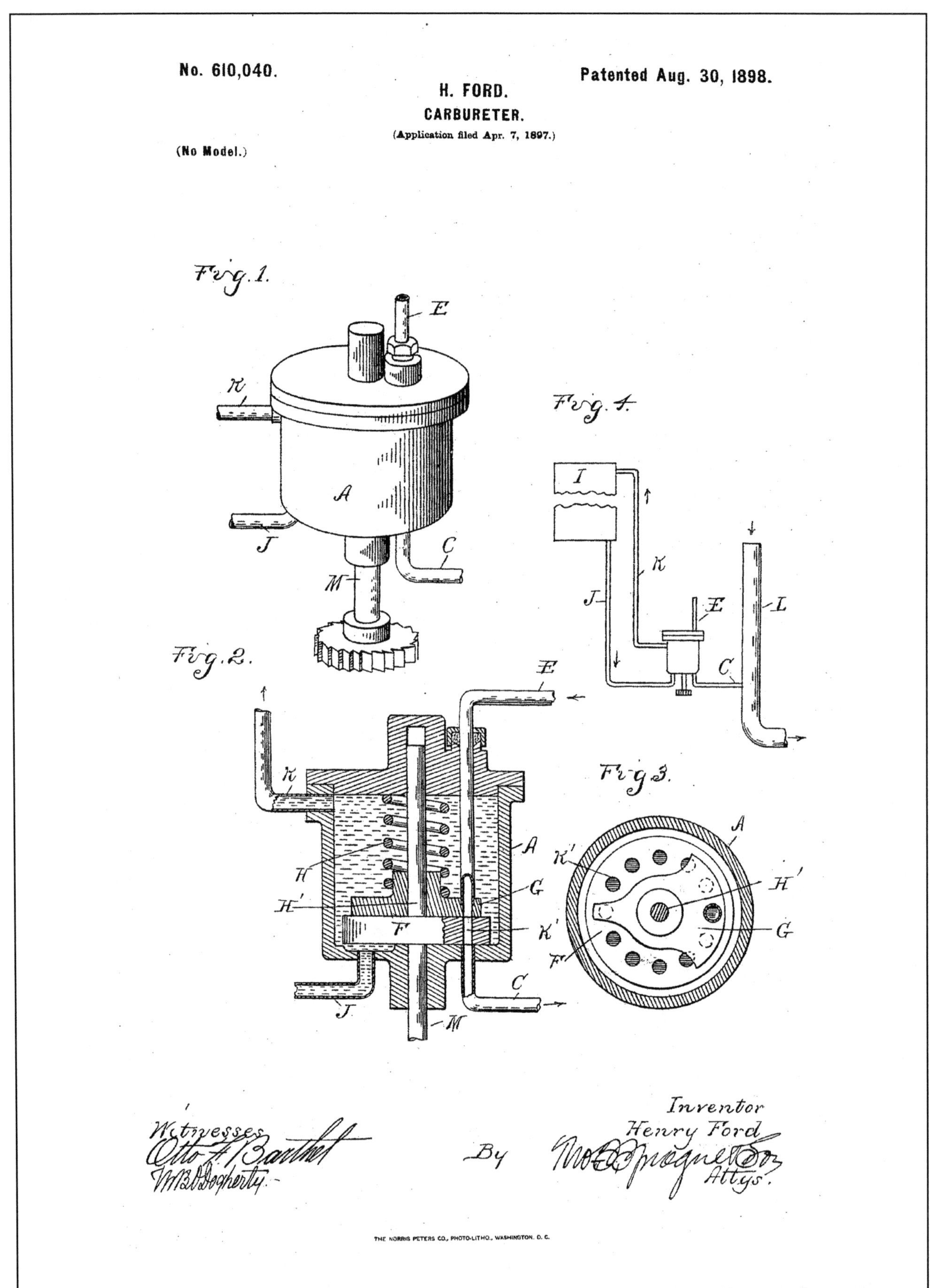

WOWidea! Move the parts not the people! Henry Ford invented many automobile components, like this one, the carburetor. He also patented a transmission and even an entirely soybean-based plastic auto body. But his real success came from creating a moving assembly process, paying decent wages and a one-size-fits-all mentality.

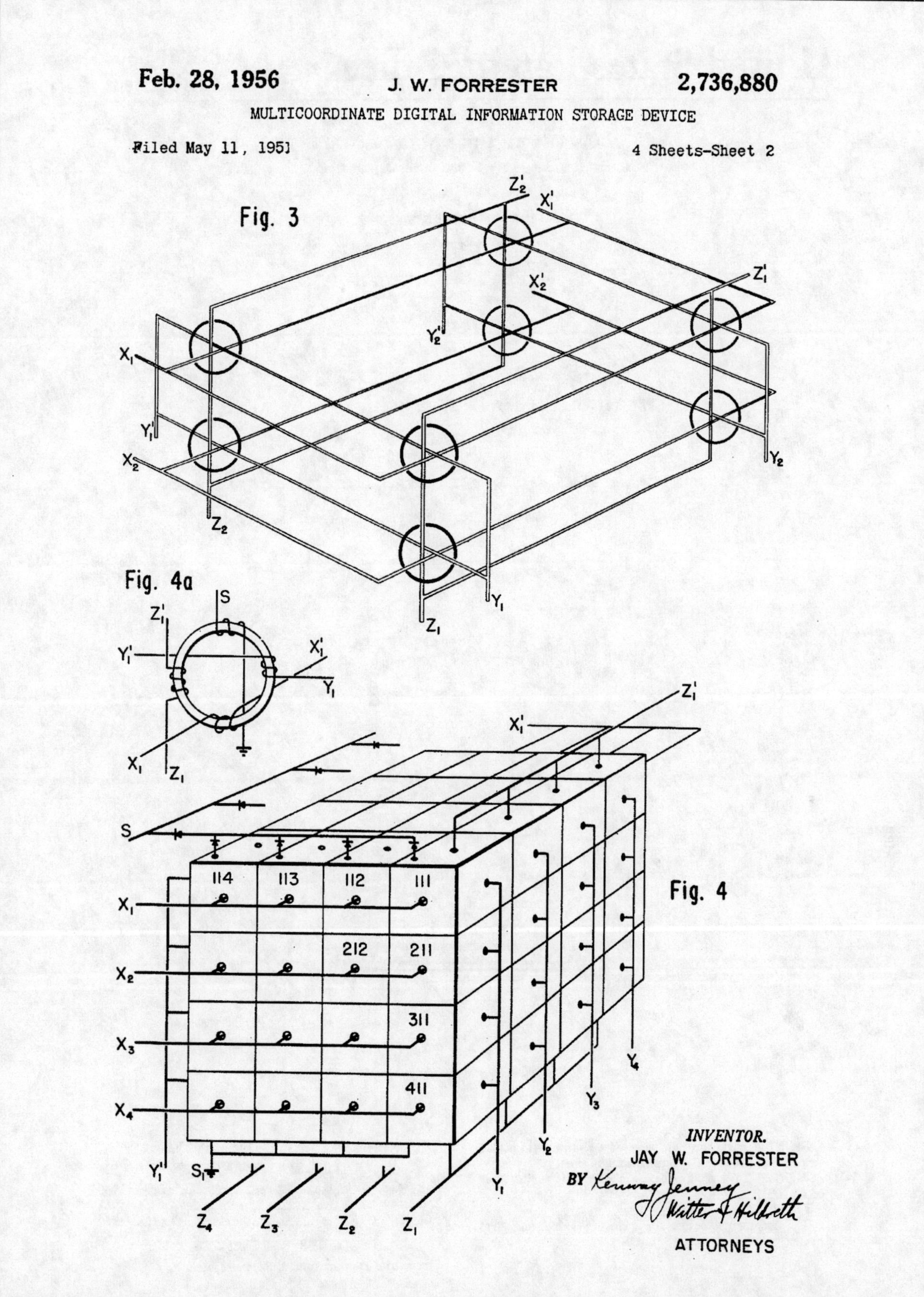

WOWidea! Flight simulator needed data to fly! Jay Forrester worked on Naval contracts to make real-time training systems. The problem was that vacuum tubes couldn't respond fast enough with the right information. Building on the work of others, his coiled matrixes of magnetized cores provided the first random access memory (RAM).

United States Patent Office

Des. 195,604
Patented July 2, 1963

195,604

CLOSURE WITH A TEAR STRIP OPENER

Ermal C. Fraze, 355 W. Stroop Road, Dayton, Ohio

Filed Feb. 4, 1963, Ser. No. 73,435

Term of patent 14 years

(Cl. D58—26)

Fig. 1

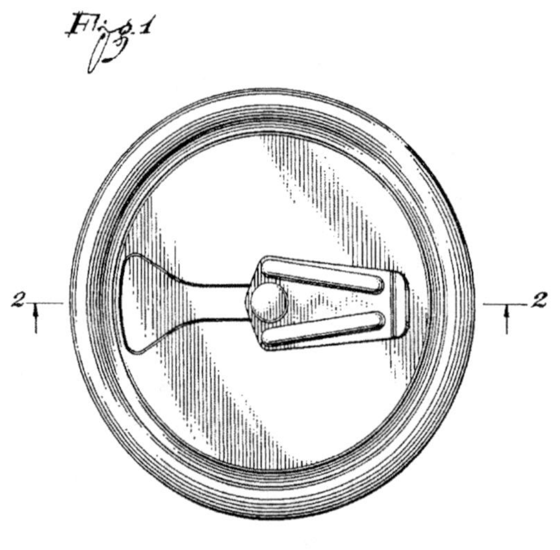

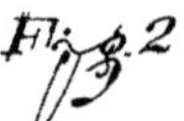

FIGURE 1 is a top plan view of a closure with a tear strip opener showing my new design; and

FIGURE 2 is a sectional view taken along the line 2—2 of FIGURE 1.

I claim:

The ornamental design for a closure with a tear strip opener, as shown.

References Cited in the file of this patent

UNITED STATES PATENTS

2,153,344	Selliken	Apr. 4, 1939
2,978,140	Walsh	Apr. 4, 1961
3,059,808	Clair	Oct. 23, 1962

OTHER REFERENCES

Industrial Design, June 1962, p. 95, item 9.

WOWidea! Attach the key to the can! Ermal Fraze couldn't sleep the night after a picnic where he desperately wanted to open a can of soda--- yet had nothing but a car bumper to attack it. 150 years after the first canning process was developed, this Ohio metal worker made the opening of the container a less dangerous act.

G. Freeman.

Wood Screws.

No. 29,963. Patented Sept. 11, 1860.

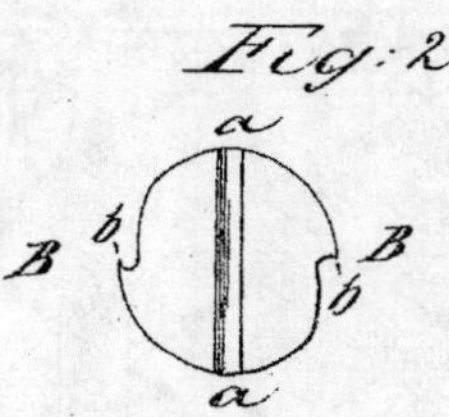

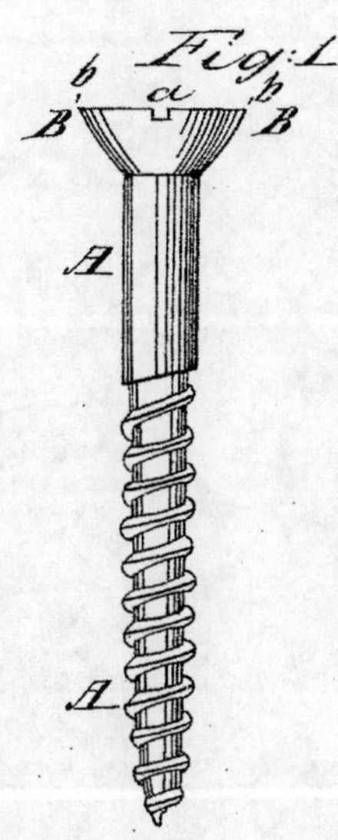

Witnesses
W. Fairfax.
C. H. Sheer

Inventor
Geo. Freeman by
A. Pollok his Atty

THE GRAPHIC CO. PHOTO.-LITH. 39 & 41 PARK PLACE, N.Y.

WOWidea! Enhancement makes history! George Freeman was not a distinguished inventor, but his version of the wood screw with its unique cutting head carries on a tradition. Archimedes, 255 BC, first created a screw to move water and from there the screw found its way into the printing press, suits of armor, carpentry and much more.

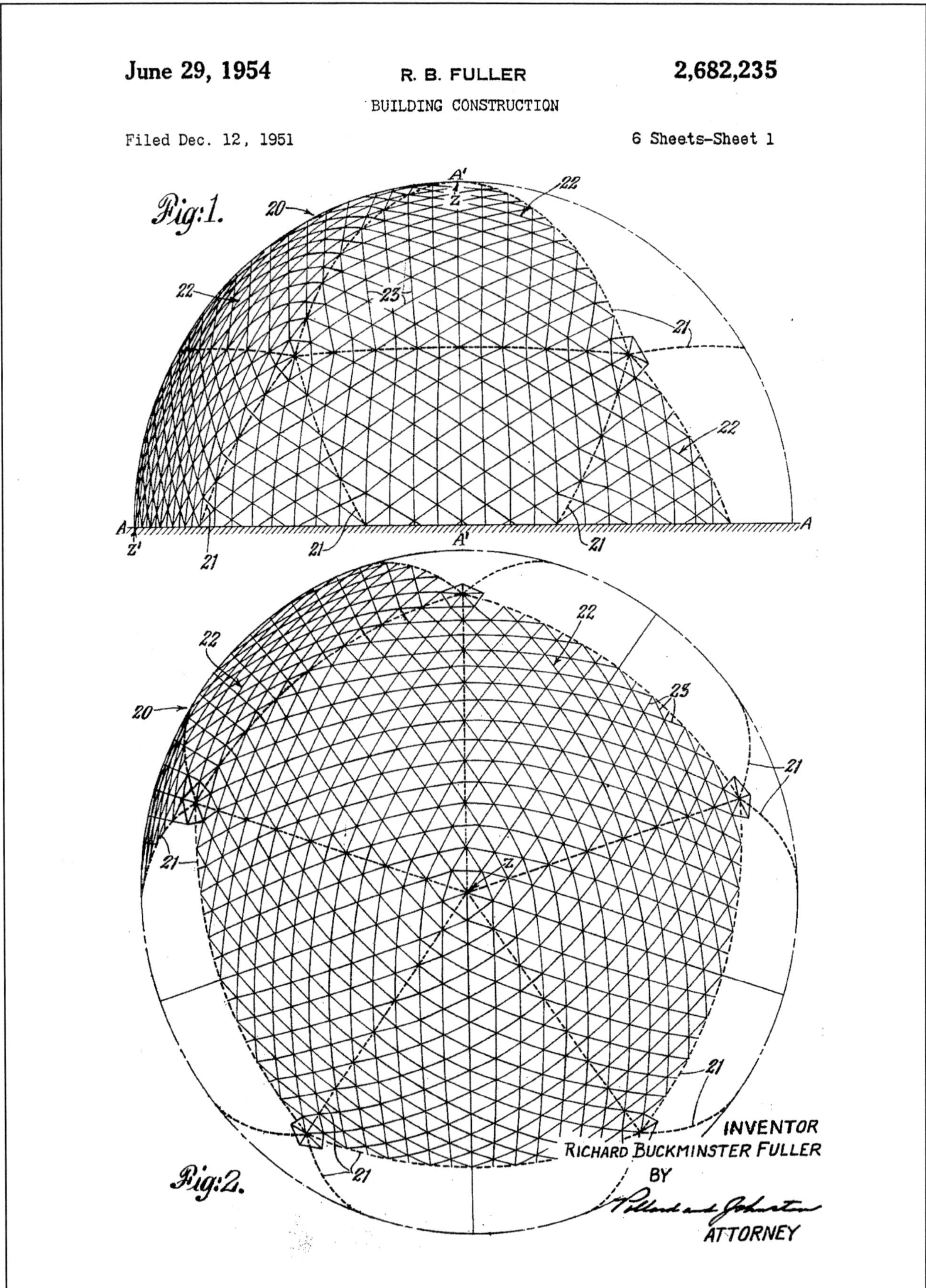

WOWidea! This dome could cover a city! Buckminster Fuller, an architect, engineer, and futurist, was asked to create assembly-line housing for aircraft workers. His work led to the self-bracing triangle design, often called the geodesic dome, it was lighter, stronger and cheaper than most every other conventional overhead enclosure.

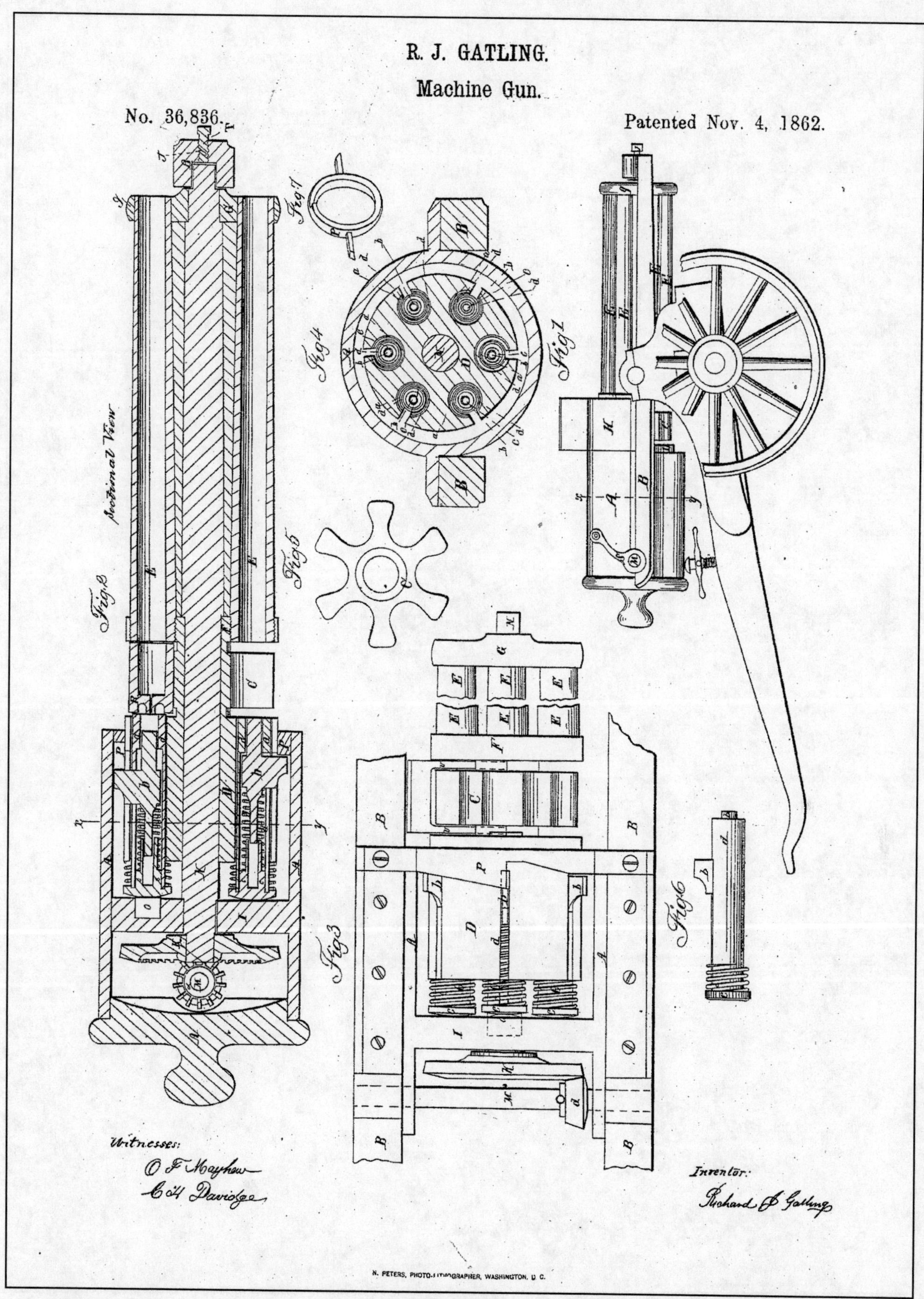

WOWidea! Man turns on man with machine! Dr. Richard Gatling was not the first to invent a mechanized gun, but his was considered to be the most "effective" design of its time. As a smart young man he patented a utilitarian rice-sowing machine and then went to medical school, yet in the end, he's known mainly for his namesake weapon.

DESIGN.

No. 37,236. PATENTED NOV. 22, 1904.

G. L. GILLESPIE.

BADGE.

APPLICATION FILED MAR. 9, 1904.

Inventor:
G. L. Gillespie

Witnesses:
E. N. Miller
Jno B. Randolph

WOWidea! Valor requires more! Civil War Brigadier General George Gillespie, received the Medal of Honor in 1867 for heroism in delivering a message (he was captured and subsequently escaped) to the Union General Sheridan. Patenting the redesign of the Army medal was meant to give it more meaning and protect it from copies.

No. 775,134. PATENTED NOV. 15, 1904.

K. C. GILLETTE.

RAZOR.

APPLICATION FILED DEC. 3, 1901.

NO MODEL.

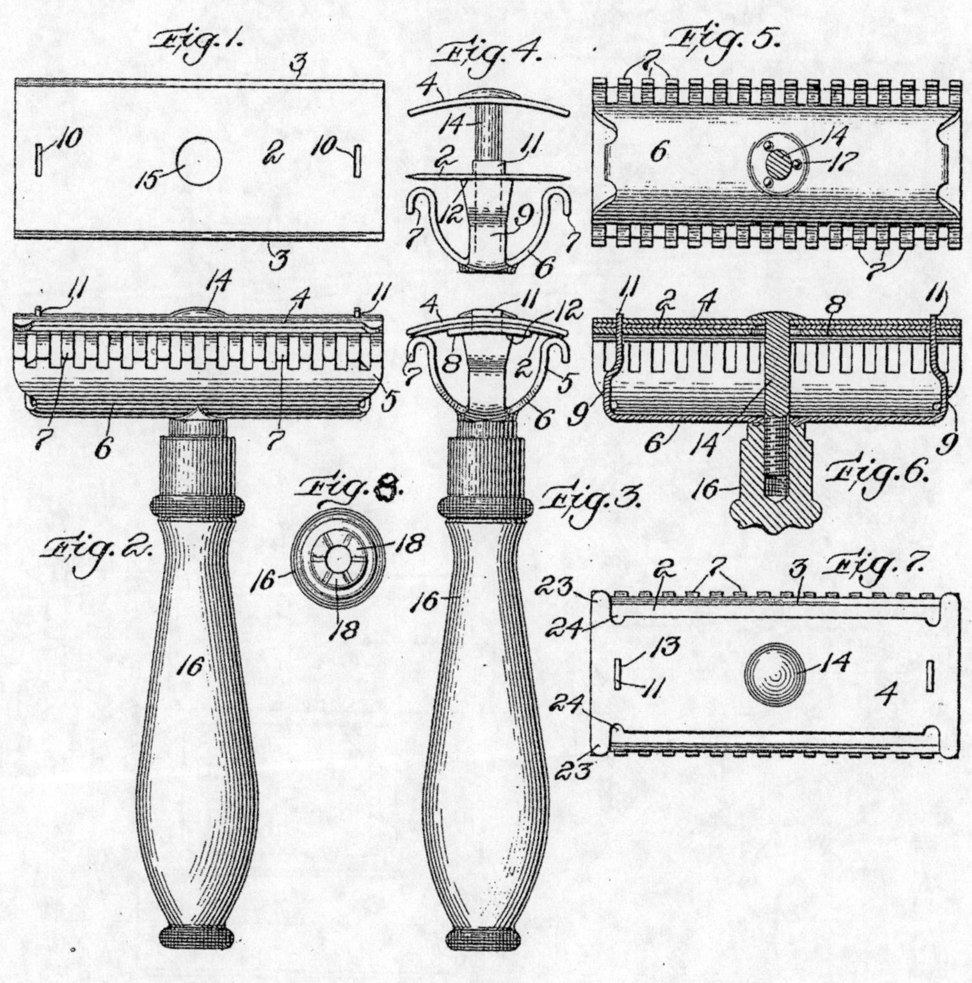

Witnesses:

Ruby M. Banfield

Malgaut A. Danihev

Inventor:

King C. Gillette,

by

E. D. Chadwick,

Attorney.

WOWidea! Lessen the nick-risk from shaving yourself! King Gillette worked for 6 years to develop the first razor with disposable steel blades and in so doing, transformed shaving from the occasional event into a daily at-home ritual. As a salesman, he had been inspired by the same man who invented the single-use cork-lined bottle cap.

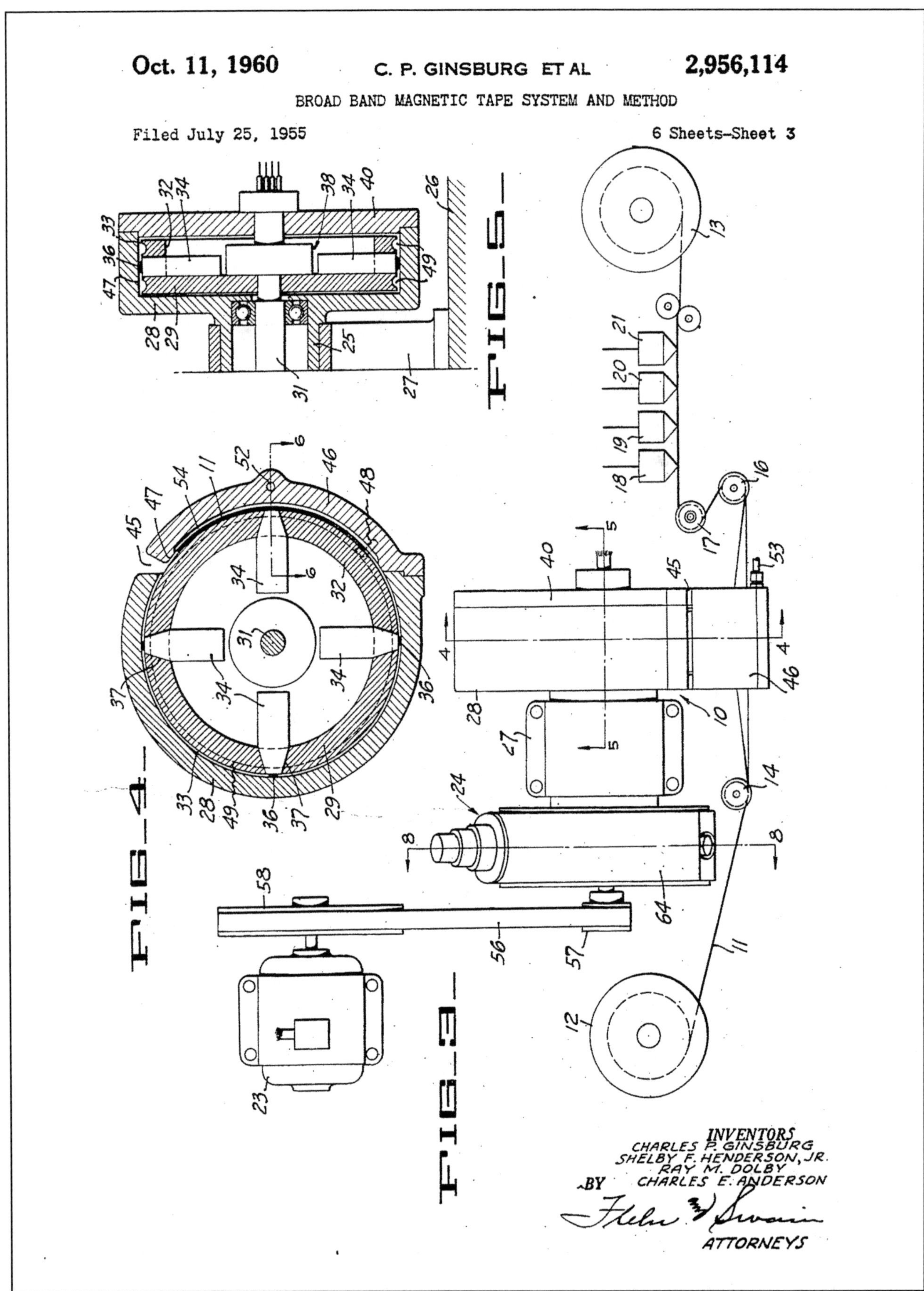

WOWidea! Turn the idea around! The first video cassette recorders were frustrating because the recording tape ran too quickly through the machine, often breaking and jamming. Charles Ginsburg, along with a team of others, devised a better way: Rotate the recording heads at high speed instead, then the tape doesn't need to run as fast.

J. F. GLIDDEN.

Wire-Fences.

No. 157,124. Patented Nov. 24, 1874.

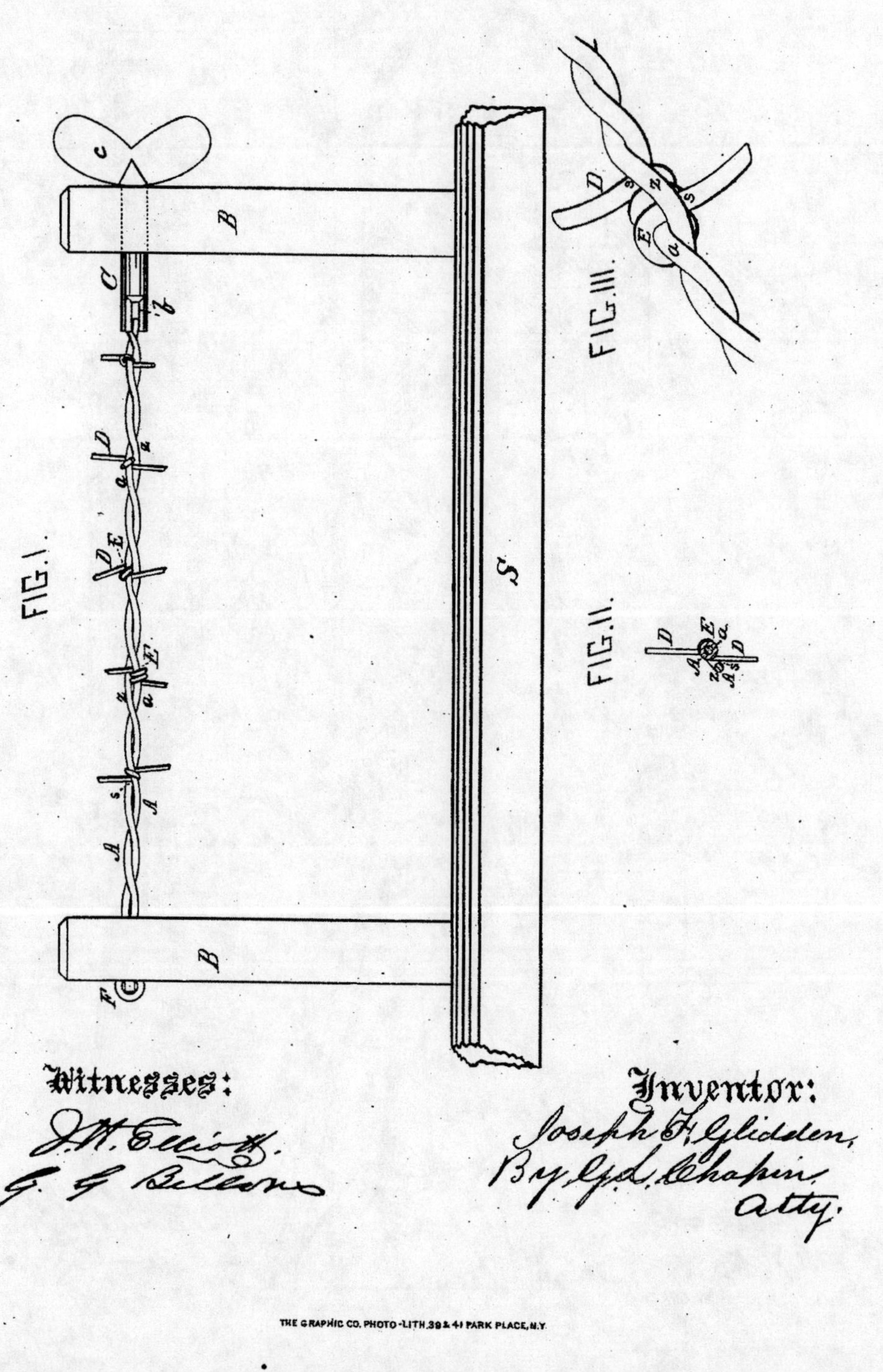

WOWidea! Good fences make good neighbors, at least on the wild frontier! Joseph Glidden, son of a farmer turned school teacher turned inventor, found his opportunity to improve on previous wired fence creations in DeKalb, Illinois. After more than 10 years of defending this patent, the Supreme Court finally affirmed it in 1891.

R. H. GODDARD.
ROCKET APPARATUS.
APPLICATION FILED OCT. 1, 1913.

1,102,653. Patented July 7, 1914.

Witnesses:
C. F. Wesson
E. J. Hartnett

Inventor
Robert H. Goddard
by attorneys
Southgate & Southgate

WOWidea! Tuberculosis provided the inventive pause! Space travel was Robert Goddard's passion since first introduced to H.G. Wells' *War of the Worlds* as a child. While at Princeton University, illness forced Goddard to rest and ultimately enabled him to submit the patent applications for his multi-stage liquid fuel rocket.

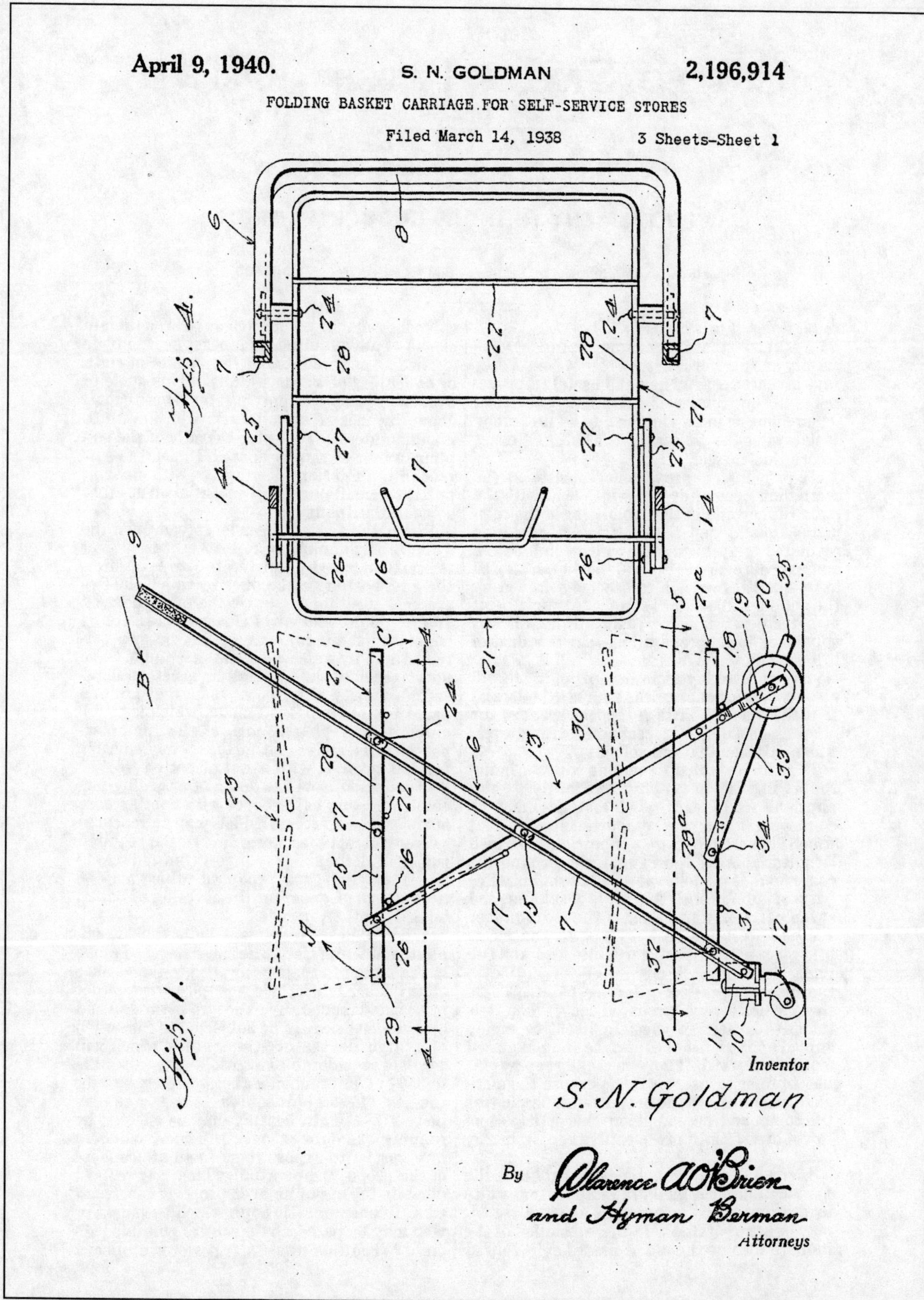

WOWidea! Push could mean profit! Woman shoppers of the day were, at first, averse to pushing anything but a baby stroller. Yet handheld baskets were heavy and could only hold so much. Long-time grocer Sylvan Goldman was onto something, so much so that by 1940 he had a 7 year waiting list for his profit-producing shopping carts.

UNITED STATES PATENT OFFICE.

CHARLES GOODYEAR, OF NEW YORK, N. Y.

IMPROVEMENT IN INDIA-RUBBER FABRICS.

Specification forming part of Letters Patent No. **3,633**, dated June 15, 1844.

To all whom it may concern:

Be it known that I, CHARLES GOODYEAR, of the city of New York, in the State of New York, have invented certain new and useful Improvements in the Manner of Preparing Fabrics of Caoutchouc or India-Rubber; and I do hereby declare that the following is a full and exact description thereof.

My principal improvement consists in the combining of sulphur and white lead with the india-rubber, and in the submitting of the compound thus formed to the action of heat at a regulated temperature, by which combination and exposure to heat it will be so far altered in its qualities as not to become softened by the action of the solar ray or of artificial heat at a temperature below that to which it was submitted in its preparation—say to a heat of 270° of Fahrenheit's scale—nor will it be injuriously affected by exposure to cold. It will also resist the action of the expressed oils, and that likewise of spirits of turpentine, or of the other essential oils at common temperatures, which oils are its usual solvents.

The articles which I combine with the india-rubber in forming my improved fabric are sulphur and white lead, which materials may be employed in varying proportions; but that which I have found to answer best, and to which it is desirable to approximate in forming the compound, is the following: I take twenty-five parts of india-rubber, five parts of sulphur, and seven parts of white lead. The india-rubber I usually dissolve in spirits of turpentine or other essential oil, and the white lead and sulphur also I grind in spirits of turpentine in the ordinary way of grinding paint. These three articles thus prepared may, when it is intended to form a sheet by itself, be evenly spread upon any smooth surface or upon glazed cloth, from which it may be readily separated; but I prefer to use for this purpose the cloth made according to the present specification, as the compound spread upon this article separates therefrom more cleanly than from any other.

Instead of dissolving the india-rubber in the manner above set forth, the sulphur and white lead, prepared by grinding as above directed, may be incorporated with the substance of the india-rubber by the aid of heated cylinders or calender-rollers, by which it may be brought into sheets of any required thickness; or it may be applied so as to adhere to the surface of cloth or of leather of various kinds. This mode of producing and of applying the sheet caoutchouc by means of rollers is well known to manufacturers. To destoy the odor of the sulphur in fabrics thus prepared, I wash the surface with a solution of potash, or with vinegar, or with a small portion of essential oil or other solvent of sulphur.

When the india-rubber is spread upon the firmer kinds of cloth or of leather it is subject to peel therefrom by a moderate degree of force, the gum letting go the fiber by which the two are held together. I have therefore devised another improvement in this manufacture by which this tendency is in a great measure corrected, and by which, also, the sheet-gum, when not attached to cloth or leather, is better adapted to a variety of purposes than when not prepared by this improved mode, which is as follows: After laying a coat of the gum, compounded as above set forth, on any suitable fabric I cover it with a bat of cotton-wool as it is delivered from the doffer of a carding-machine, and this bat I cover with another coat of the gum—a process which may be repeated two or three times, according to the required thickness of the goods. A very thin and strong fabric may be thus produced, which may be used in lieu of paper for the covering of boxes, books, or other articles.

When this compound of india-rubber, sulphur, and white lead, whether to be used alone in the state of sheets or applied to the surface of any other fabric, has been fully dried, either in a heated room or by exposure to the sun and air, the goods are to be subjected to the action of a high degree of temperature, which will admit of considerable variation—say from 212° to 350° of Fahrenheit's thermometer, but for the best effect approaching as nearly as may be to 270°. This heating may be effected by running the fabrics over a heated cylinder; but I prefer to expose them to an atmosphere of the proper temperature, which may be best done by the aid of an oven properly constructed with openings through which the sheet or web may be passed by means of suitable rollers. When this process is performed upon a

WOWidea! The rubber industry was melting and heat was the fix! Natural rubber, derived from trees in Central America became soft in summer and brittle in winter. It was up to Charles Goodyear, a man in and out of debtors prison, to discover on a stove by accident, that the secret was to add sulfur and heat the gooey mixture to cure it.

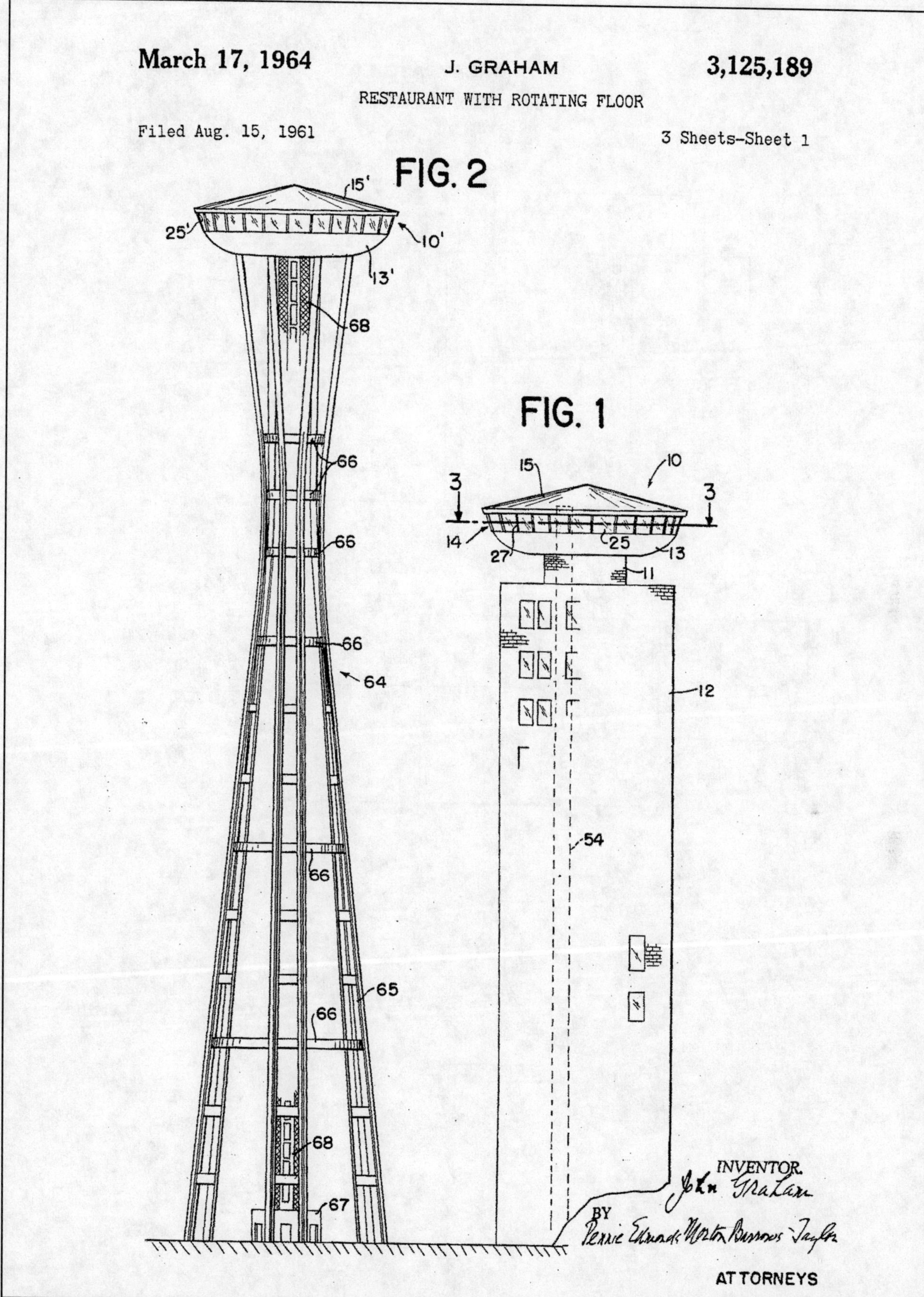

WOWidea! Revolving restaurant and tourist attraction all-in-one! Conceived as part of the 1960 World's Fair in Seattle, Washington, USA, John Graham was an architect like his father. He created, among other things, this downtown Space Needle® and a suburban shopping center, which was to become the basis for the modern mall.

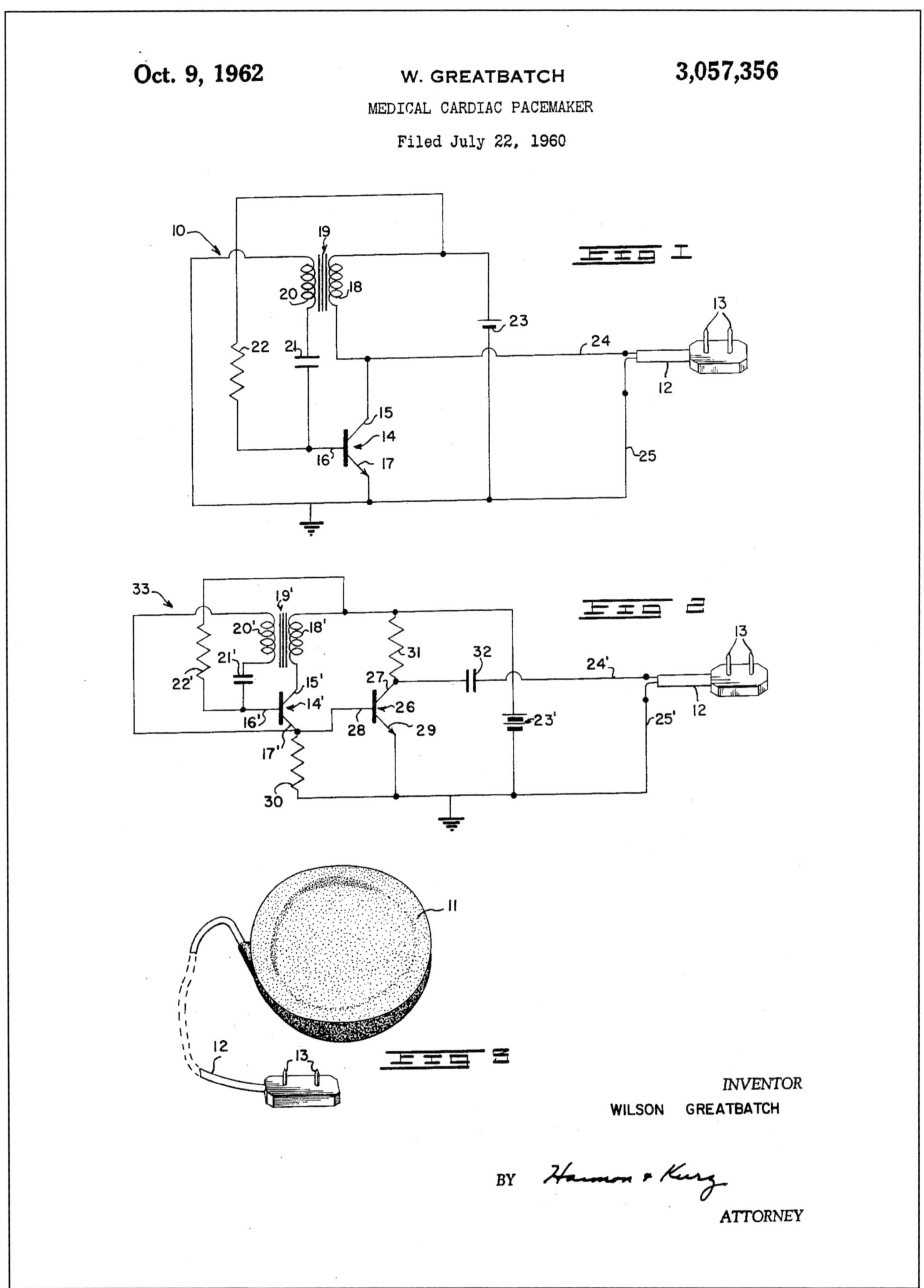

WOWidea! Jump-start the heart from inside! Until Wilson Greatbatch brought the Cold War's technology boom to use in preventing pulmonary failure, doctors were using external cardiac pulse generators the size of a toaster. The real life-saving benefit in this design was that, instead of hooking it up *after* a heart attack, it was always on-duty.

C. GREENWOOD.

EAR-MUFFLERS.

No. 188,292. Patented March 13, 1877.

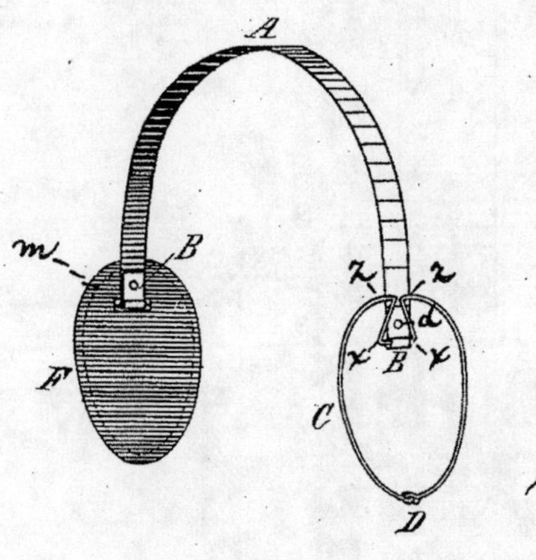

Fig.1.

Fig.2.

Witnesses:

Lewel G. Shaw

H. E. Metcalf

Inventor

Chester Greenwood,

Per C. C. Shaw

Atty.

N. PETERS, PHOTO-LITHOGRAPHER, WASHINGTON, D.C.

WOWidea! Necessity was the mother of his invention! The cold-winter months in Maine provided warmth to an idea when Chester Carlson, a 19 year-old boy, ventured outside to ice skate and nearly froze his ears. He went on to manufacture up to 400,000 ear muffs a year and was awarded over 100 patents during his lifetime.

(No Model.)

C. M. HALL.

MANUFACTURE OF ALUMINIUM.

No. 400,665. Patented Apr. 2, 1889.

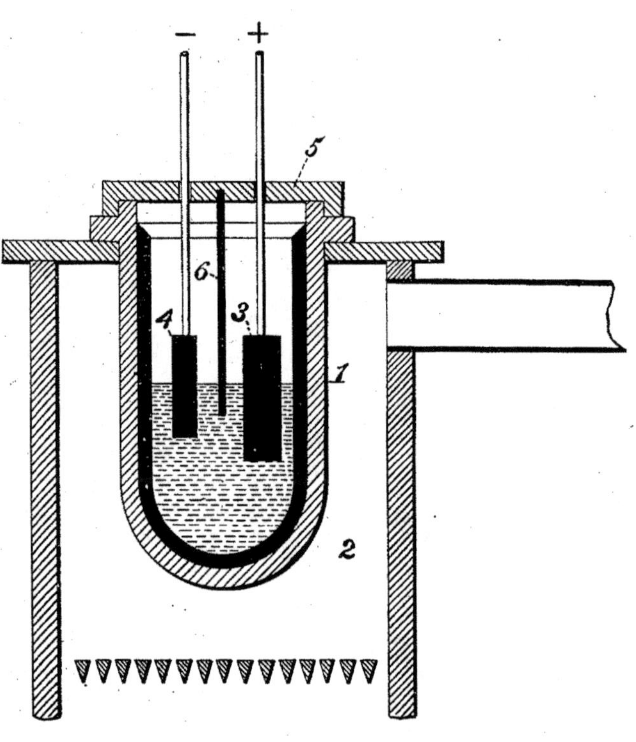

WITNESSES:

E. Newell.

F. E. Gaither

INVENTOR,

Charles M. Hall

by Darwin S. Wolcott. Att'y.

WOWidea! Changing the crucible from clay to carbon was crucial! Many had previously produced aluminum, but Hall, along with his sister Julia, experimented endlessly to uncover the right combination of elements to create a practical method for mass production. His patent was challenged, but upheld by then-Judge William Howard Taft.

United States Patent [19]

Head

[11] **3,999,756**

[45] **Dec. 28, 1976**

[54] **TENNIS RACKET**

[75] Inventor: **Howard Head,** Baltimore, Md.

[73] Assignee: **Prince Manufacturing, Inc.,** Princeton, N.J.

[22] Filed: **Sept. 10, 1975**

[21] Appl. No.: **612,076**

Related U.S. Application Data

[63] Continuation-in-part of Ser. No. 516,550, Oct. 21, 1974, abandoned.

[52] **U.S. Cl.** .... **273/73 C;** 273/73 D
[51] **Int. Cl.²** **A63B 49/02**
[58] **Field of Search** .... 273/73 R, 73 C, 73 D, 273/73 F, 73 H, 29 A, 67 R

[56] **References Cited**

UNITED STATES PATENTS

801,246	10/1905	Johnson	273/73 D
1,539,019	5/1925	Nikonow	273/73 R
2,164,631	7/1939	Abell	273/73 H
3,086,777	4/1963	Lacoste	273/73 H
3,305,235	2/1967	Williams	273/73 CX
3,515,386	6/1970	Mason	273/73 R
3,545,756	12/1970	Nash	273/73 D
3,801,099	4/1974	Lair	273/73 C
3,820,785	6/1974	Occhipinti et al.	273/73 D X
3,917,267	11/1975	McGrath	273/73 C

FOREIGN PATENTS OR APPLICATIONS

178,843	6/1954	Austria	273/73 C
848,826	8/1970	Canada	273/73 C
427,206	4/1935	United Kingdom	273/73 R
755,257	8/1956	United Kingdom	273/73 R

Primary Examiner—Richard J. Apley
Attorney, Agent, or Firm—Seidel, Gonda & Goldhammer

[57] **ABSTRACT**

The tennis racket has a strung surface which is larger than the strung surface of a conventional racket, particularly in regard to its dimension in a longitudinal direction from the frame tip toward the handle shaft of the racket. The conventional length, weight, and balance which have proven necessary for good playing characteristics for all tennis rackets of the past have been maintained. The racket has unexpectedly achieved increased strength and a combination of advantages in playing characteristics without resort to weights, springs, or other complications previously proposed. The racket has a zone of high coefficient of restitution, much larger than that of conventional rackets, extending in a longitudinal direction from the region of the center of percussion to a point 1¼ inch from the throat of the racket, thereby taking maximum advantage of the location of the center of percussion of the racket. The zone of high coefficient of restitution is also wider with respect to the corresponding zone on conventional rackets.

18 Claims, 19 Drawing Figures

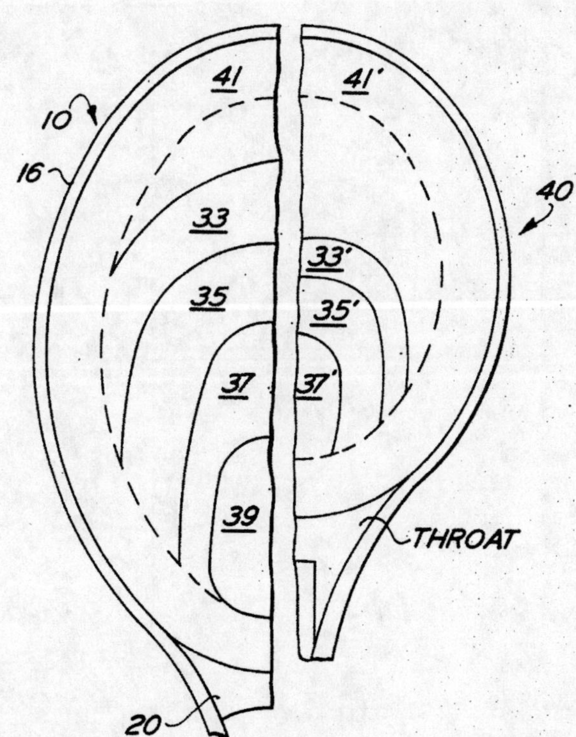

WOWidea! Expanded sweet spot falls just short of banishment! Howard Head, already a legendary ski innovator, wanted to improve his chances for victory on the courts too. This led to his simple reshaping (there were no rule-book definitions at the time) of the traditional racket, which, in turn, led to greater enjoyment of the game for many.

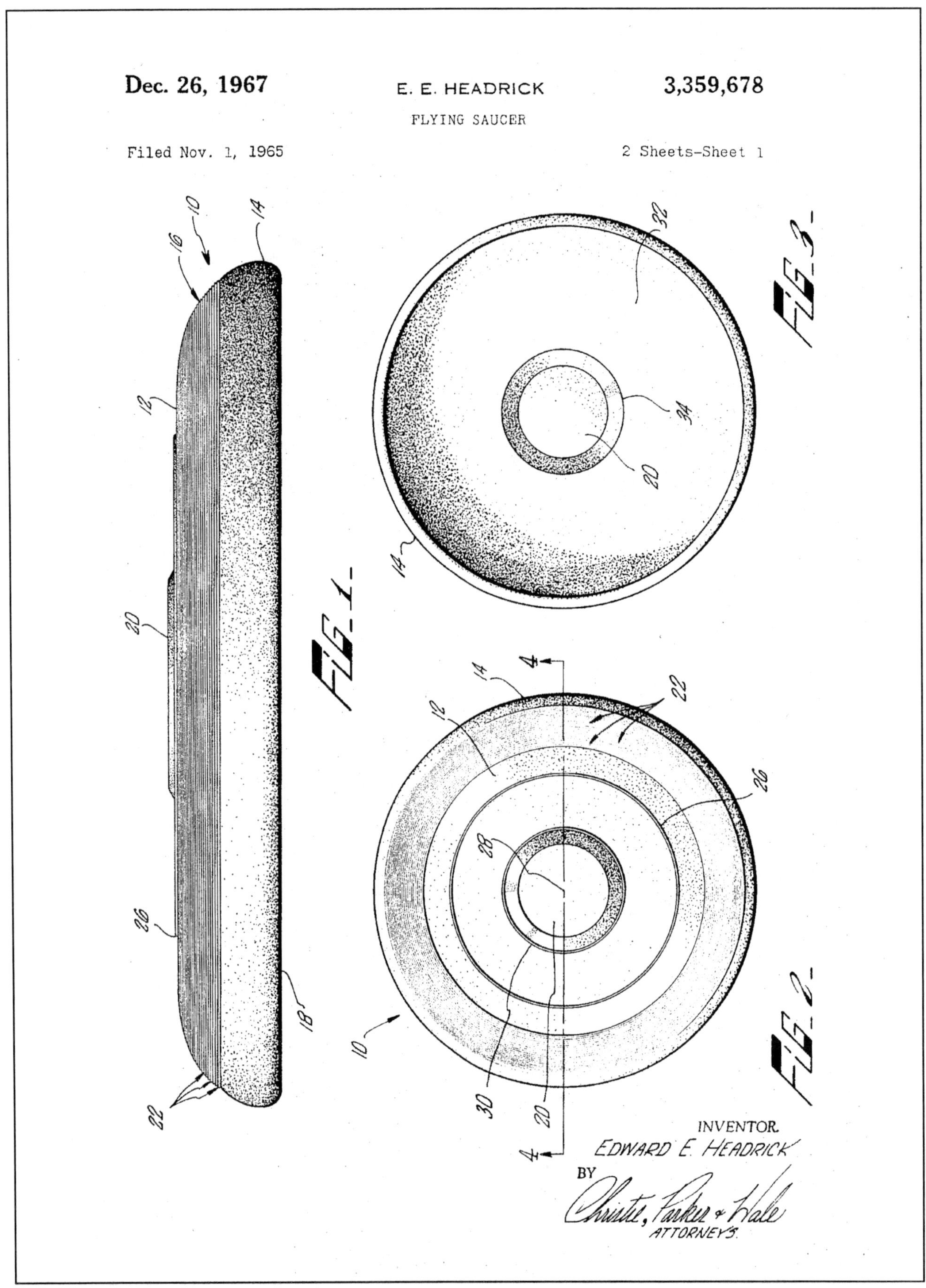

WOWidea! Baked goods lead to recreational icon! During the late 1800s, the Olds Bakery Company employed a William Fresbie and so named their pies after him. Soon, nearby college students learned to fly the empty pie-tins. Years later, Edward Headrick came along to forever immortalize the pastime with his playful plastic version.

2 Sheets—Sheet 1.

J. H. HEINZ.
Vegetable Assorters.

No. 212,000. Patented Feb. 4, 1879.

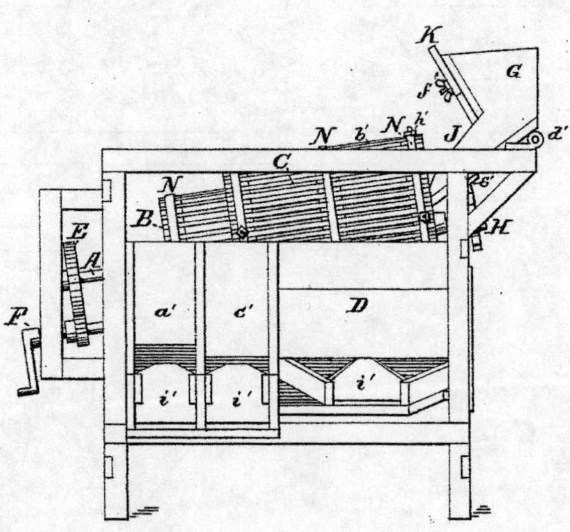

Fig 1.

WITNESSES:

John V. Torrance
Jno F. Ullwood

John H. Heinz.
INVENTOR

per John B. Geyser.
ATTORNEY

N. PETERS, PHOTO-LITHOGRAPHER, WASHINGTON, D. C.

WOWidea! The king of ketchup created more than 57 varieties of success! After two unfounded arrests for fraud and one financial bankruptcy (due to his commitment to purchase an unforeseen bumper crop of cucumbers and cabbage in 1875), Heinz rose from those setbacks to take his pickled vegetables to a huge and hungry market.

Jan. 6, 1942. W. R. HEWLETT 2,268,872

VARIABLE FREQUENCY OSCILLATION GENERATOR

Filed July 11, 1939

FIG.1.

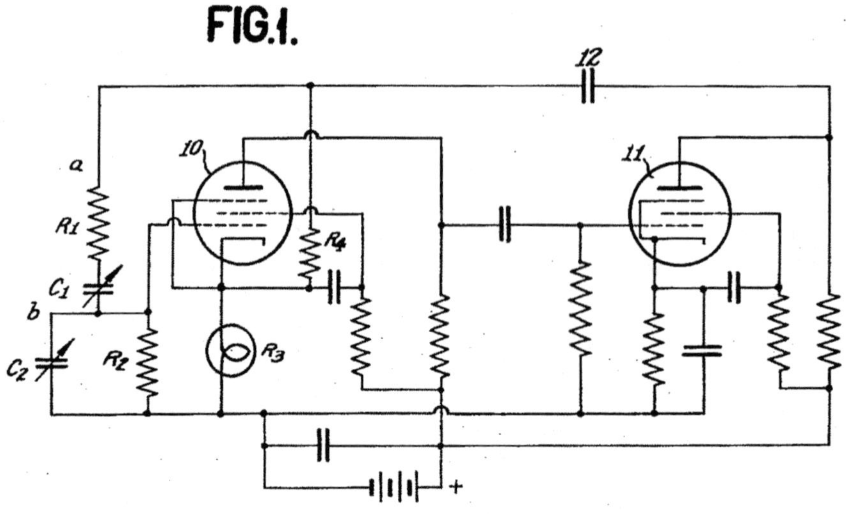

FIG. 2.

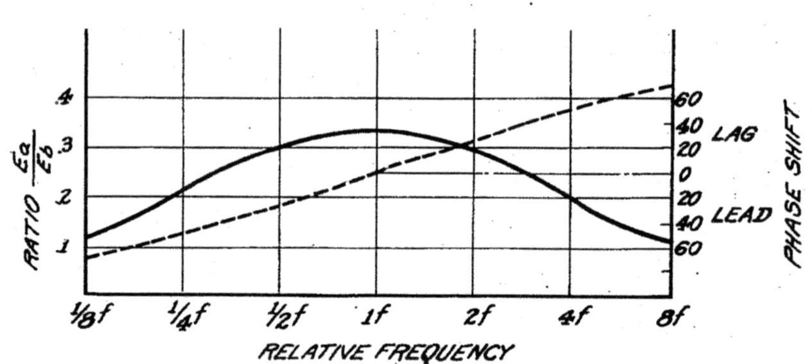

INVENTOR
WILLIAM R. HEWLETT
BY Paul D. Flehr
ATTORNEY

WOWidea! Simple little light bulb made the sound smooth! William Hewlett, eventual co-founder of Hewlett-Packard, first created an inexpensive pure-tone audio generator as part of his thesis work at Stanford University. Once on the market, Walt Disney Studios was first to use the H-P Model 200A in their creation of Fantasia®.

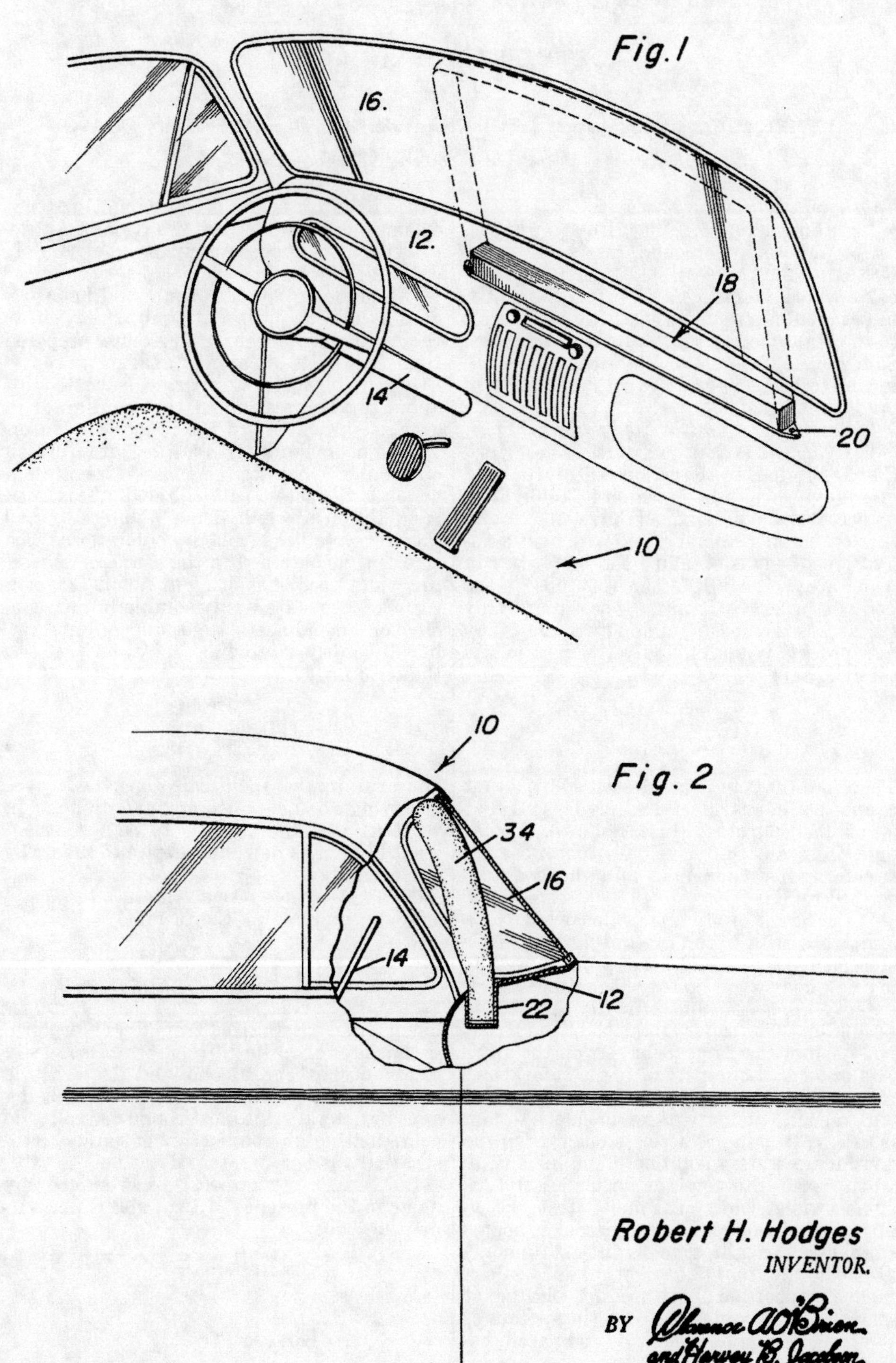

WOWidea! Airbags came before 3-point seatbelts! This forward-looking invention was likely overlooked by the automotive industry for years because the technology of its day wasn't able fulfill its promise. Robert Hodges was never able to make it a commercial success, but without this patent, vehicle safety might have remained in neutral.

UNITED STATES PATENT OFFICE.

FELIX HOFFMANN, OF ELBERFELD, GERMANY, ASSIGNOR TO THE FARBEN-FABRIKEN OF ELBERFELD COMPANY, OF NEW YORK.

ACETYL SALICYLIC ACID.

SPECIFICATION forming part of Letters Patent No. 644,077, dated February 27, 1900.

Application filed August 1, 1898. Serial No. 687,385. (Specimens.)

To all whom it may concern:

Be it known that I, FELIX HOFFMANN, doctor of philosophy, chemist, (assignor to the FARBENFABRIKEN OF ELBERFELD COMPANY, of New York,) residing at Elberfeld, Germany, have invented a new and useful Improvement in the Manufacture or Production of Acetyl Salicylic Acid; and I hereby declare the following to be a clear and exact description of my invention.

In the *Annalen der Chemie und Pharmacie*, Vol. 150, pages 11 and 12, Kraut has described that he obtained by the action of acetyl chlorid on salicylic acid a body which he thought to be acetyl salicylic acid. I have now found that on heating salicylic acid with acetic anhydride a body is obtained the properties of which are perfectly different from those of the body described by Kraut. According to my researches the body obtained by means of my new process is undoubtedly the real acetyl salicylic acid

$$C_6H_4\left\langle\begin{matrix}OCO.CH_3\\COOH.\end{matrix}\right.$$

Therefore the compound described by Kraut cannot be the real acetyl salicylic acid, but is another compound. In the following I point out specifically the principal differences between my new compound and the body described by Kraut.

If the Kraut product is boiled even for a long while with water, (according to Kraut's statement,) acetic acid is not produced, while my new body when boiled with water is readily split up, acetic and salicylic acid being produced. The watery solution of the Kraut body shows the same behavior on the addition of a small quantity of ferric chlorid as a watery solution of salicylic acid when mixed with a small quantity of ferric chlorid—that is to say, it assumes a violet color. On the contrary, a watery solution of my new body when mixed with ferric chlorid does not assume a violet color. If a melted test portion of the Kraut body is allowed to cool, it begins to solidify (according to Kraut's statement) at from 118° to 118.5° centigrade, while a melted test portion of my product solidifies at about 70° centigrade. The melting-points of the two compounds cannot be compared, because Kraut does not give the melting-point of his compound. It follows from these details that the two compounds are absolutely different.

In producing my new compound I can proceed as follows, (without limiting myself to the particulars given:) A mixture prepared from fifty parts of salicylic acid and seventy-five parts of acetic anhydride is heated for about two hours at about 150° centigrade in a vessel provided with a reflux condenser. Thus a clear liquid is obtained, from which on cooling a crystalline mass is separated, which is the acetyl salicylic acid. It is freed from the acetic anhydride by pressing and then recrystallized from dry chloroform. The acid is thus obtained in the shape of glittering white needles melting at about 135° centigrade, which are easily soluble in benzene, alcohol, glacial acetic acid, and chloroform, but difficultly soluble in cold water. It has the formula

$$C_6H_4\left\langle\begin{matrix}OCOCH_3\\COOH\end{matrix}\right.$$

and exhibits therapeutical properties.

Having now described my invention and in what manner the same is to be performed, what I claim as new, and desire to secure by Letters Patent, is—

As a new article of manufacture the acetyl salicylic acid having the formula:

$$C_6H_4\left\langle\begin{matrix}O.COCH_3\\COOH\end{matrix}\right.$$

being when crystallized from dry chloroform in the shape of white glittering needles, easily soluble in benzene, alcohol and glacial acetic acid, difficultly soluble in cold water, being split by hot water into acetic acid and salicylic acid, melting at about 135° centigrade, substantially as hereinbefore described.

In testimony whereof I have signed my name in the presence of two subscribing witnesses.

FELIX HOFFMANN.

Witnesses:
R. E. JAHN,
OTTO KÖNIG.

WOWidea! Greek physician Hippocrates prescribed the bark and leaves of the willow tree (rich in a substance called salicin) to relieve pain and fever! 2000 years later, Felix Hoffmann, at Bayer® in Germany, chemically synthesized a stable powder form that relieved his father's rheumatism. They named the compound aspirin.

(No Model.) 3 Sheets—Sheet 1.

H. HOLLERITH.

ART OF COMPILING STATISTICS.

No. 395,782. Patented Jan. 8, 1889.

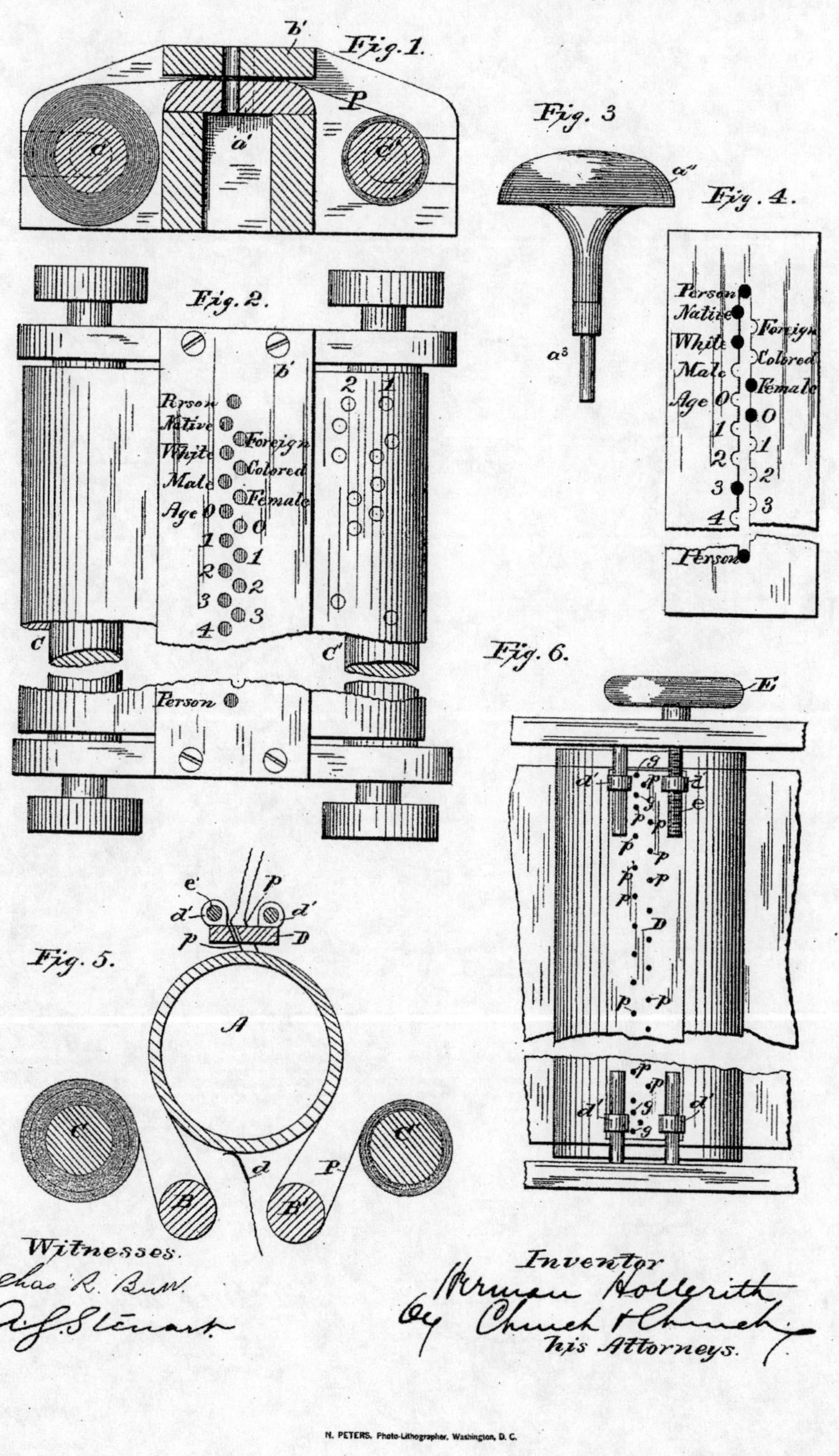

WOWidea! A train was the ticket! The U.S. census took years to complete in the 1800s because everything was counted by hand. Herman Hollerith borrowed a conductor's method, which kept track of who you were with holes punched in a card, to solve the census crisis, creating an early computing machine and resulting company, IBM®

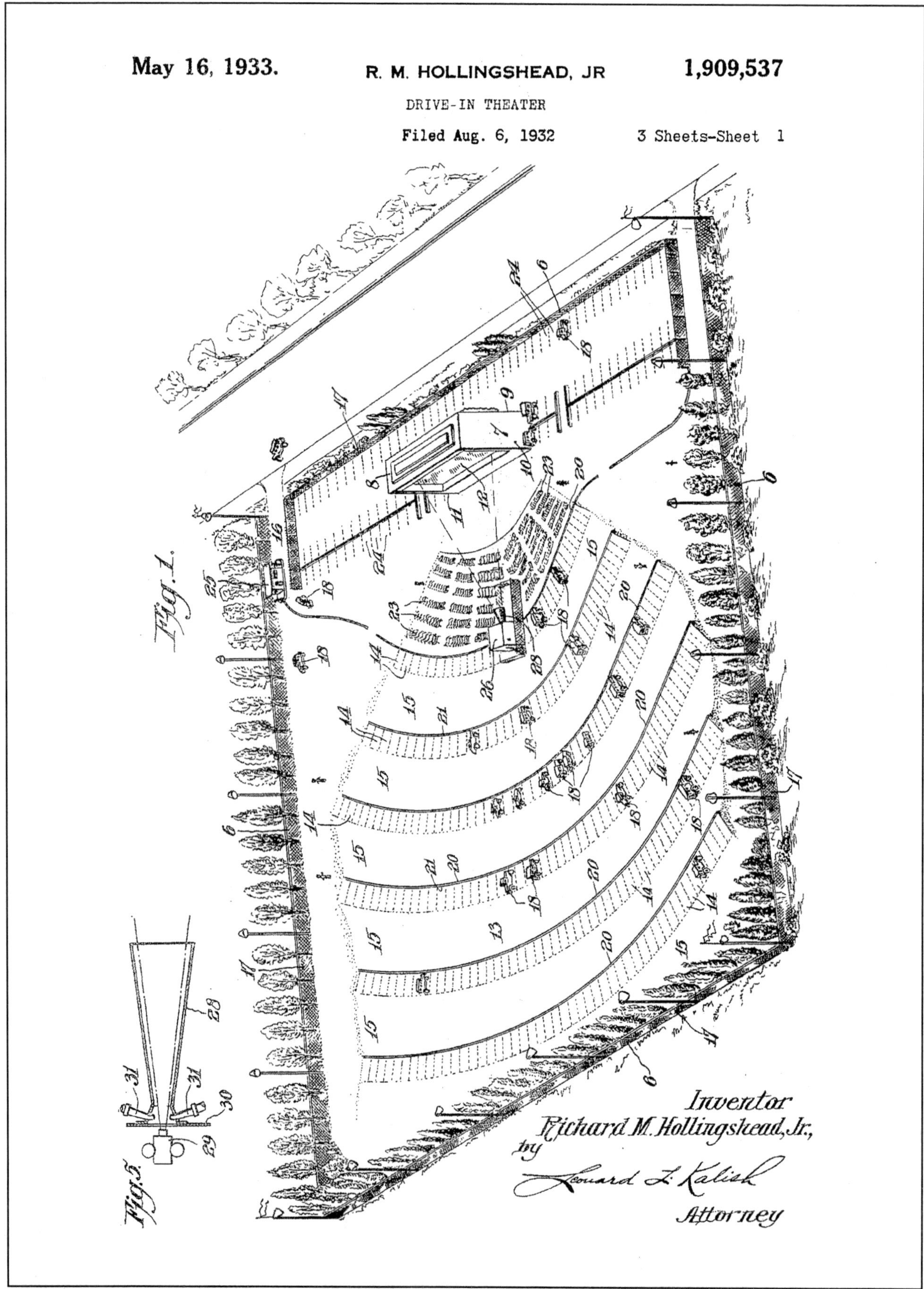

WOWidea! "Sit in your car, see and hear movies!" Trumpeting a new era in entertainment, the patent was ultimately declared invalid by the courts 17 years later, but the drive-in theater was born following Richard Hollingshead's trials at his New Jersey home, where films were projected on a screen nailed to the inventor's backyard trees.

(No Model.)

W. C. HOOKER.

ANIMAL TRAP.

No. 528,671. Patented Nov. 6, 1894.

Fig. 1.

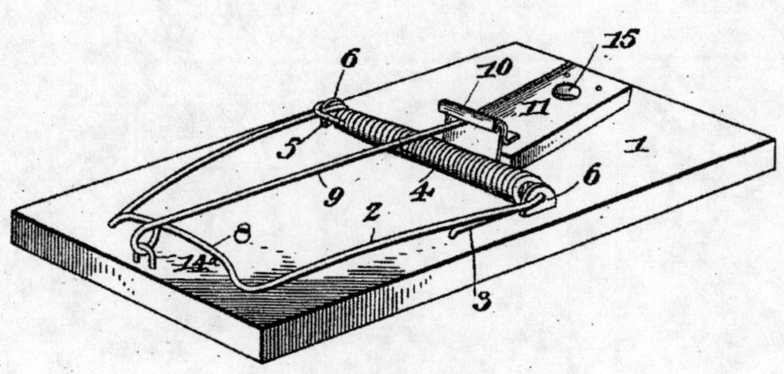

Fig. 2.

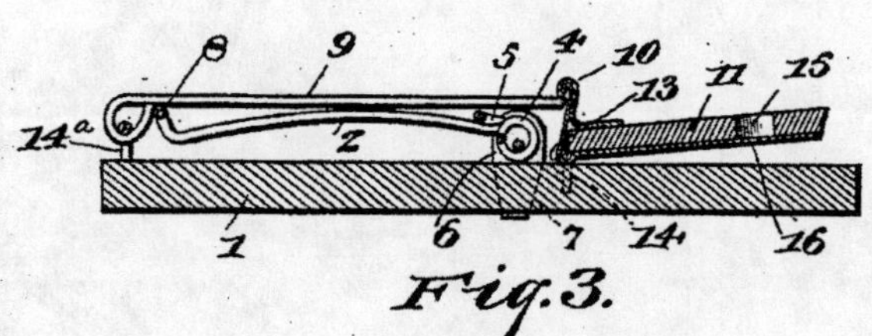

Fig. 3.

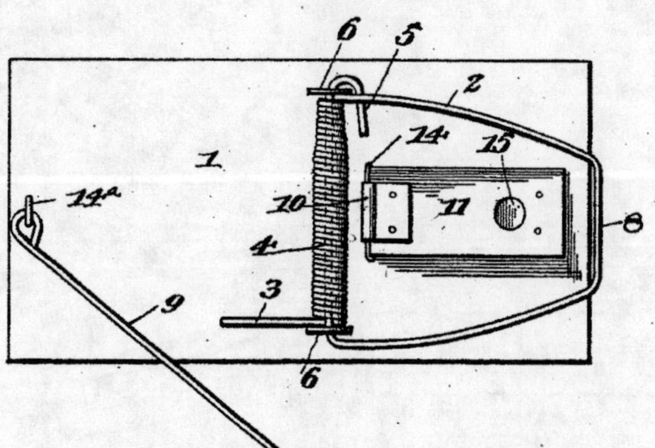

Witnesses
B. S. Ober
T. F. Riley

Inventor
William C. Hooker,

By his Attorneys.
C. A. Snow & Co.

WOWidea! He built a better mouse trap! Considered the first easy-to-set mouse trap of its kind, William Hooker kept on trying to improve his animal catcher but, with 27 patents to his name, this one proved the best at snapping up "customers". Still manufactured, it proves that simple yet effective designs have tremendous staying power.

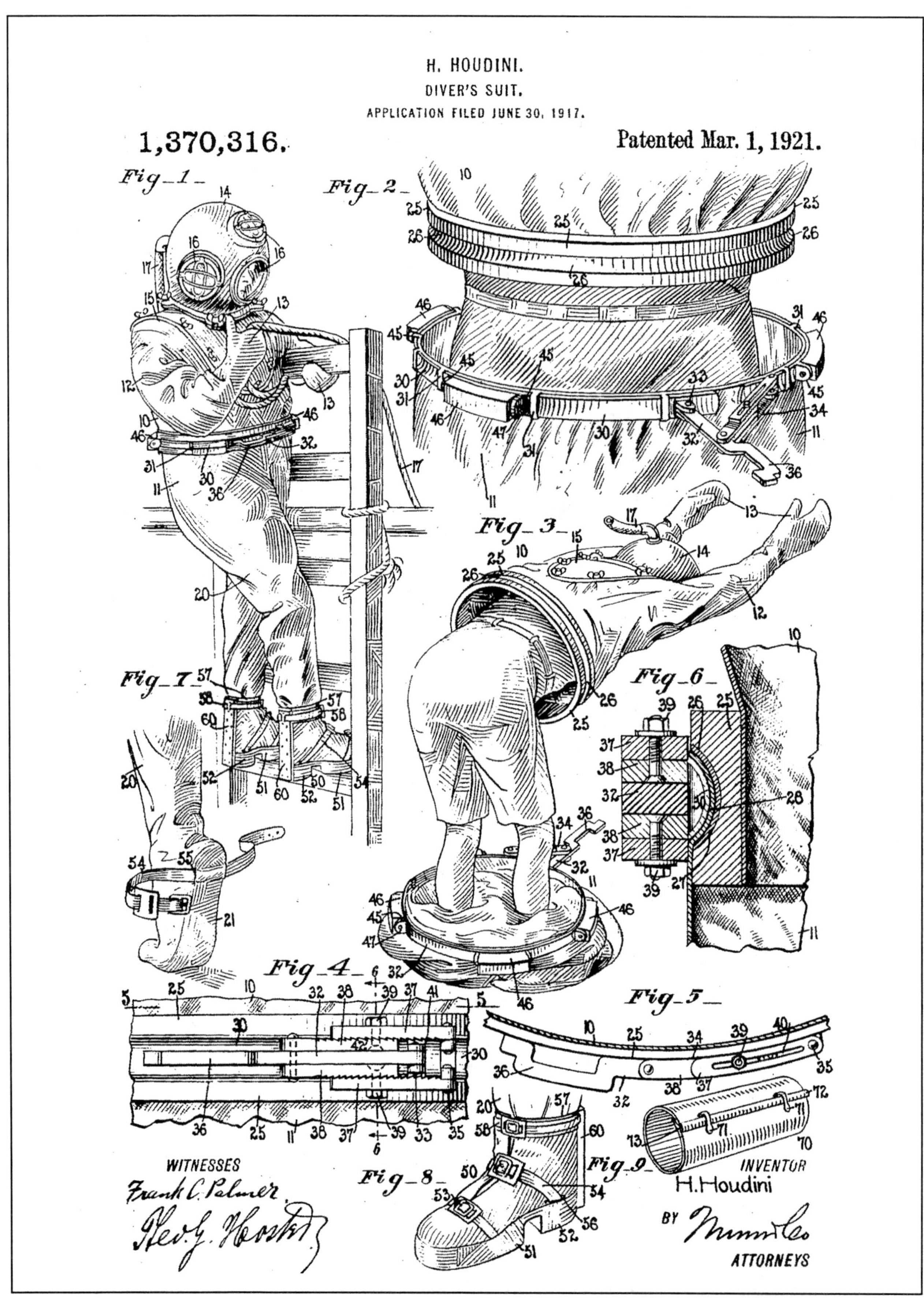

WOWidea! Safety suit works like magic! Harry Houdini was no ordinary magician, in addition to the wondrous shows Houdini publicly performed, he was committed to revealing frauds by unscrupulous charlatans and mentalists of the day. This magic trick turned divers' release mechanism was just another way he helped others.

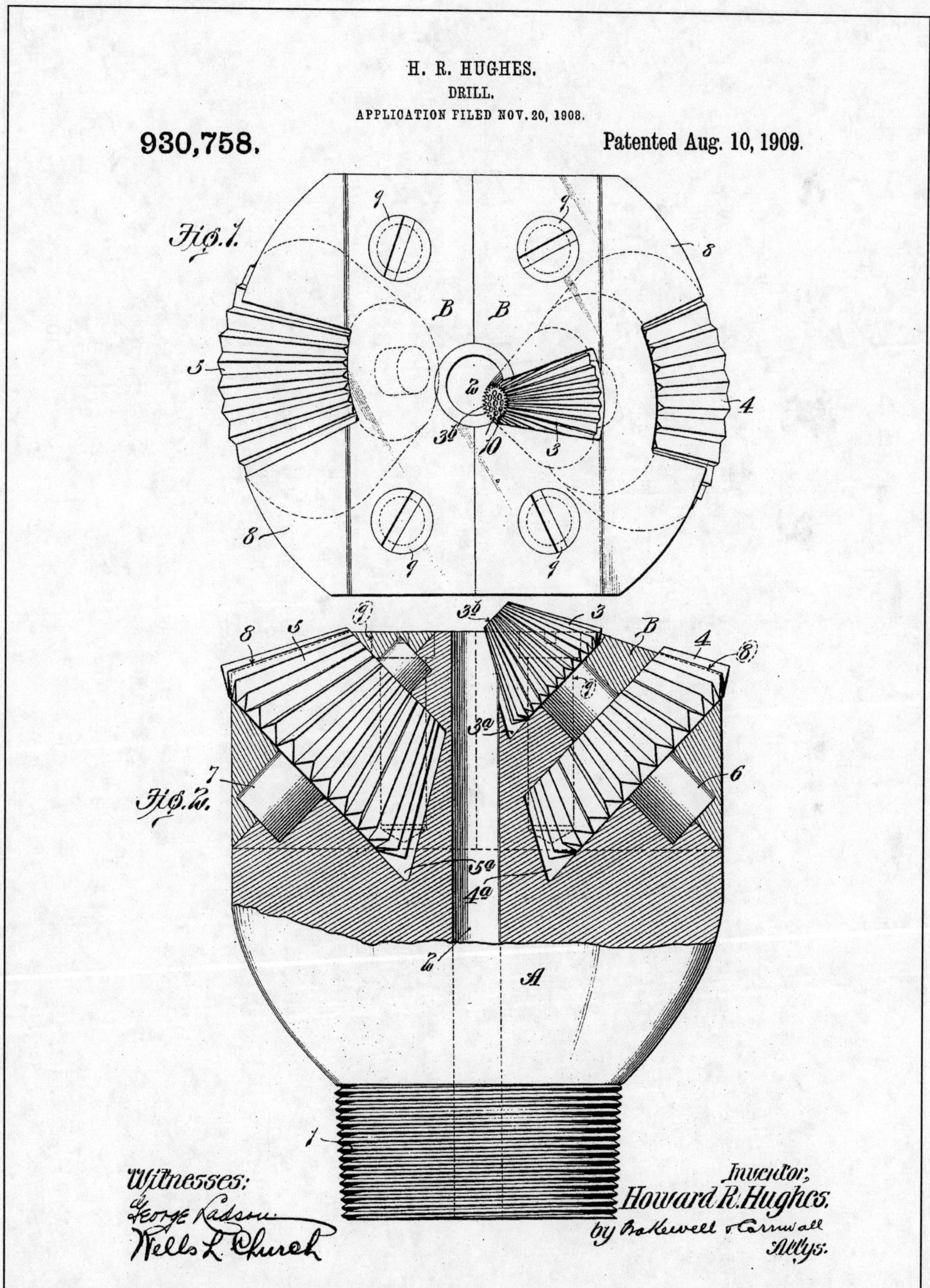

WOWidea! Instead of wearing down the rock, it pulverized it! Howard Hughes Sr. was drilling for oil when he discovered much more. This one idea would some day make his son the richest man in the world--- providing enough wealth to spur the growth of the aircraft and movie industries and even the gambling mecca of Las Vegas.

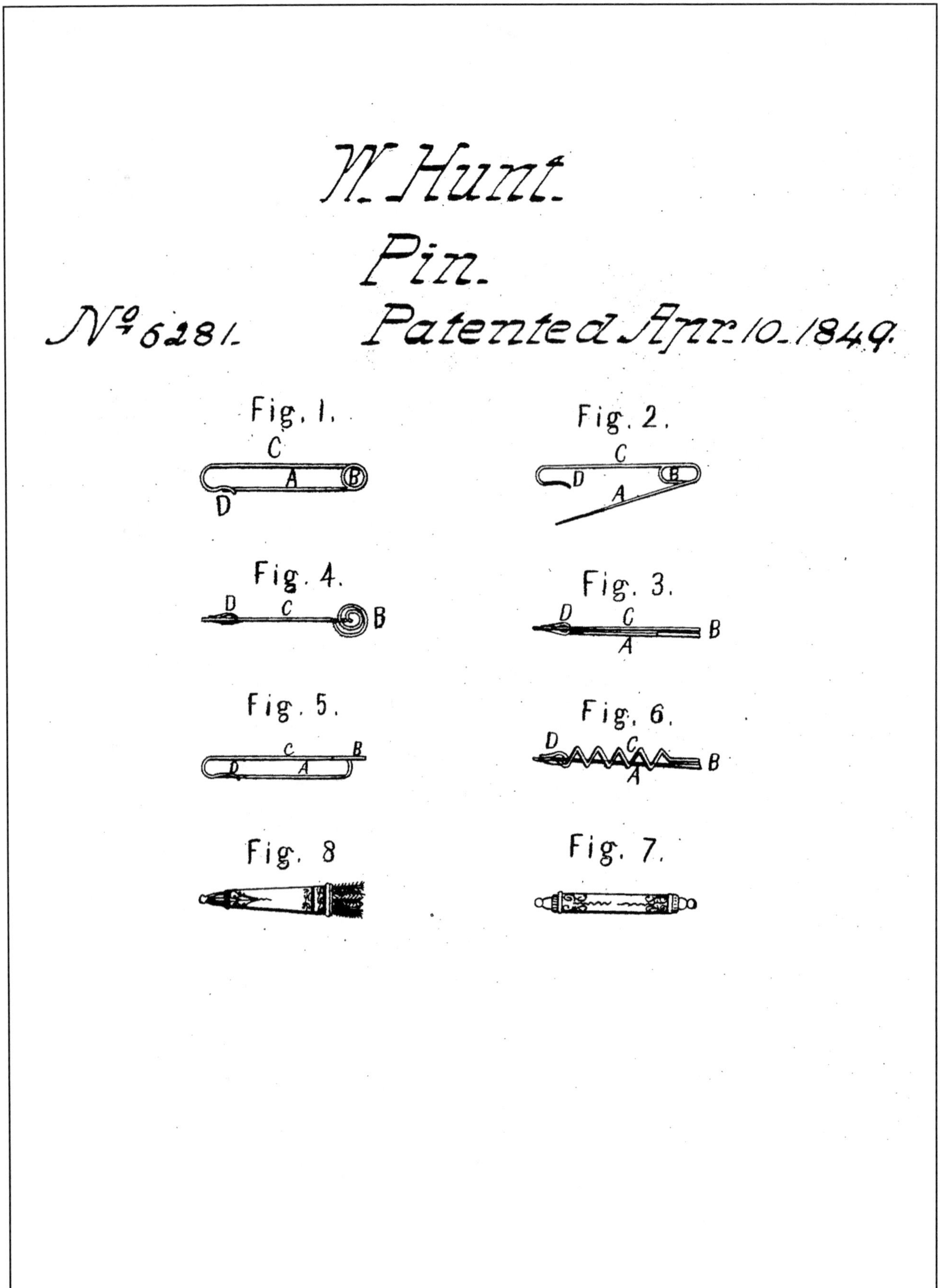

W. Hunt.

Pin.

No. 6281. Patented Apr. 10. 1849.

Fig. 1.
C
A
B
D

Fig. 2.
C
D
B
A

Fig. 4.
D C B

Fig. 3.
D C B
A

Fig. 5.
C B
D A

Fig. 6.
D C B
A

Fig. 8

Fig. 7.

WOWidea! For lack of funds he kept us from being undone! William Hunt, a steady inventor with a number of patents already under his belt, owed money to his patent draftsmen, the Richardsons. On a bet, they would forgive his debt, by gaining the rights to any device he created with a piece of wire, 3 hours later--- the "safety" pin.

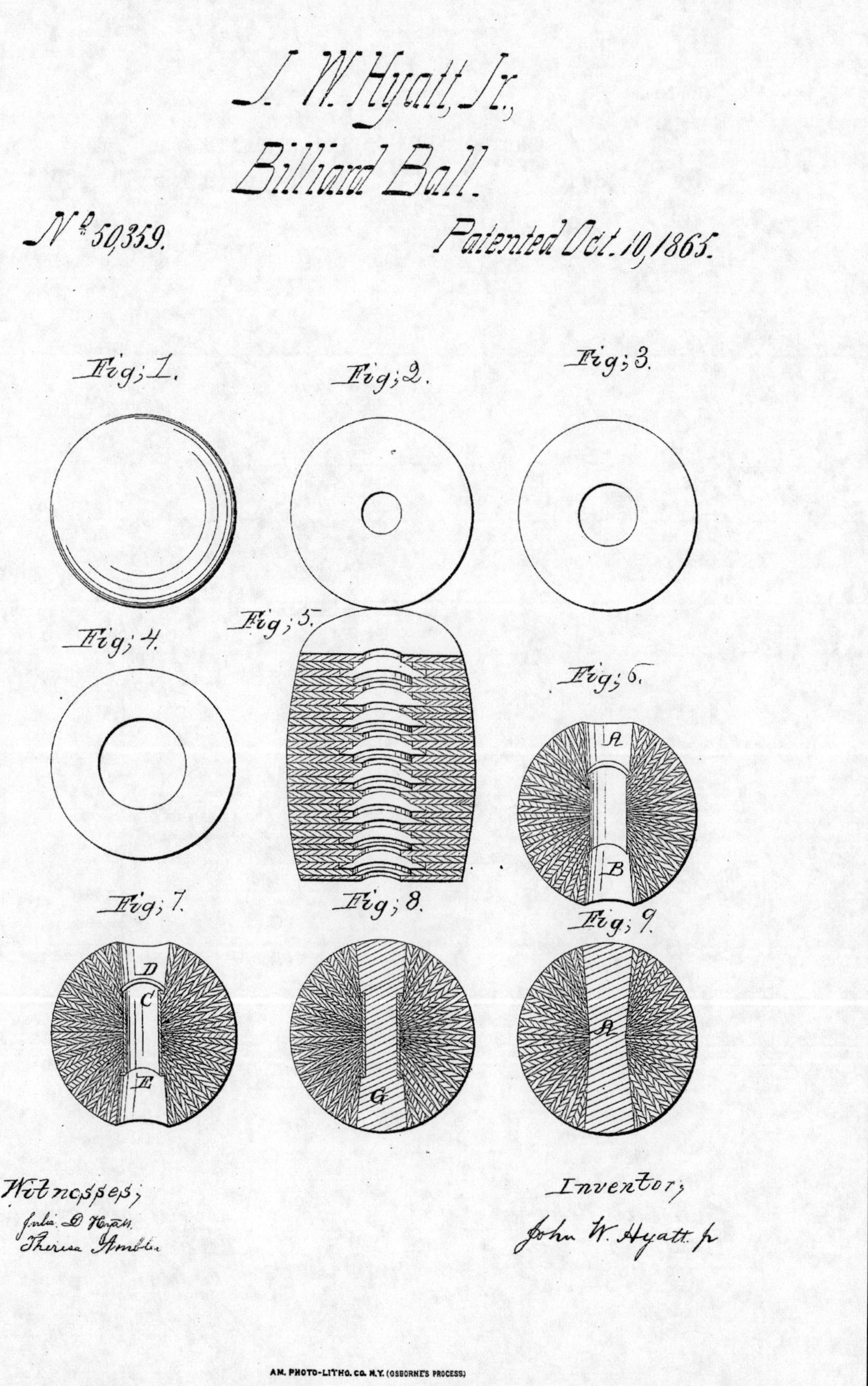

WOWidea! Survival of the species credited to contest! The Civil War-era popularity of billiards forced the slaughter of hundreds of elephants to obtain their ivory for balls. John Hyatt, noticing a $10,000 reward for the first person to find a suitable substitute, won. Celluloid, also a Hyatt invention, gave birth to modern day plastics.

(No Model.)

F. E. IVES.

PHOTOGRAVURE PRINTING PLATE.

No. 495,341. Patented Apr. 11, 1893.

FIG. 1.

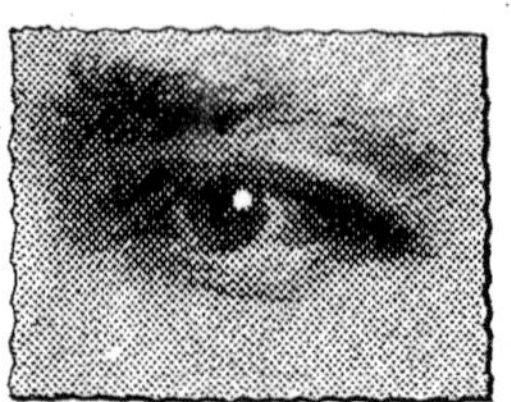

a a

Witnesses:
Alex. Barkoff
A. V. Groupe

Inventor:
Frederic E. Ives
by his Attorneys
Howson & Howson

WOWidea! Individual dots make image appear whole! Following an apprenticeship, Frederic Ives traveled to Philadelphia to create the first halftone method of printing photographs. A gelatin relief was created from the negative and pressed against a "grid" of crossed lines. The resulting dots were transferred to a plate for printing.

US005552982A

United States Patent [19]

Jackson et al.

[11] **Patent Number: 5,552,982**

[45] **Date of Patent: Sep. 3, 1996**

[54] **METHOD AND SYSTEM FOR PROCESSING FIELDS IN A DOCUMENT PROCESSOR**

[75] Inventors: **Peter C. Jackson**, Seattle; **William H. Gates, III**, Redmond; **Bryan Loofbourrow**, Seattle, all of Wash.

[73] Assignee: **Microsoft Corporation**, Redmond, Wash.

[21] Appl. No.: **205,141**

[22] Filed: **Feb. 28, 1994**

Related U.S. Application Data

[63] Continuation of Ser. No. 694,001, Apr. 30, 1991, abandoned, which is a continuation of Ser. No. 607,872, Oct. 31, 1990, abandoned.

[51] **Int. Cl.**[6] **G06F 17/22**

[52] **U.S. Cl.** **364/419.1**; 364/419.17; 395/145

[58] **Field of Search** 364/419.19, 419.14, 364/419.17, 419.1; 395/146, 600, 147, 148, 145

[56] **References Cited**

U.S. PATENT DOCUMENTS

4,965,763	10/1990	Zamora	364/419.19
5,025,396	6/1991	Parks et al.	395/147
5,148,366	9/1992	Buchanan et al.	395/146

OTHER PUBLICATIONS

"Using Microsoft Word", Word Processing Program Version 3.0 ©Microsoft Corporation '86. pp. 259–273.

"Parallel Computations on Strings & Arrays", Crochemore et al ©1990 Springer–Verlag, Germany.

"Numeric Processor and Text Manipulator for the 'Master Contro' Data Base Management System", Energy Research & Den't Admin. Feb. 1976.

"Using Microsoft Word" Version 3 ©1983, 1985, 1986. pp. 259–273.

"Switch" definition, *Microsoft Press–Computer Dictionary,* second edition, copyright 1994, p. 378.

"Switch" and Switch Code definitions, *Dictionary of Computers, Information Processing, and Telecommunications,* second edition, copyright 1984, pp. 617–618.

Primary Examiner—Robert A. Weinhardt
Attorney, Agent, or Firm—Seed and Berry LLP

[57] **ABSTRACT**

A method and system for providing calculated fields in a document processing environment. A calculated field comprises a beginning field character, a field code, and an ending field character. The field code specifies a behavior of the calculated field. A user may insert a calculated field into a document. After insertion, the result of the calculated field can be determined based on the behavior specified by the field code. The result of the calculated field is then stored within the document and preferably within the calculated field. The document can be displayed in either field code or results mode. In field code mode, when the document is displayed, the calculated field is displayed (i.e., beginning field character, field code, and ending field character). In results mode, when the document is displayed, the result is displayed, rather than the calculated field.

27 Claims, 15 Drawing Sheets

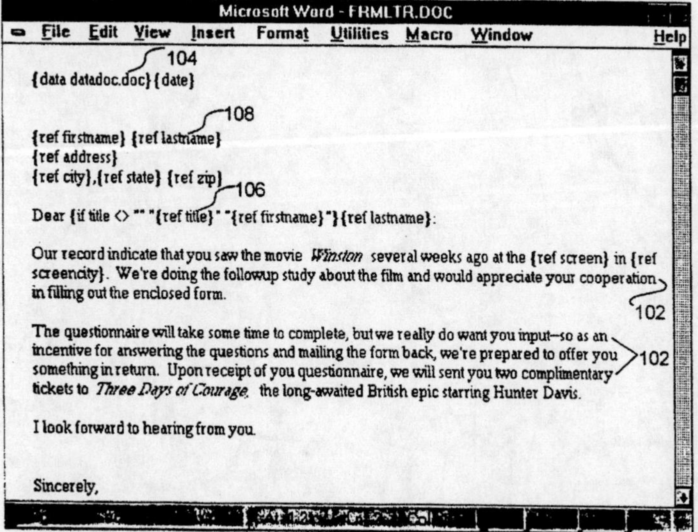

WOWidea! Mail merge makes marketers merry! Although he had some help from two colleagues, Bill Gates is no slouch when it comes to outside-the-post-office-box thinking. Their concept of using software codes within a computer-based document to grab and insert specific information changed direct mail campaigns forever.

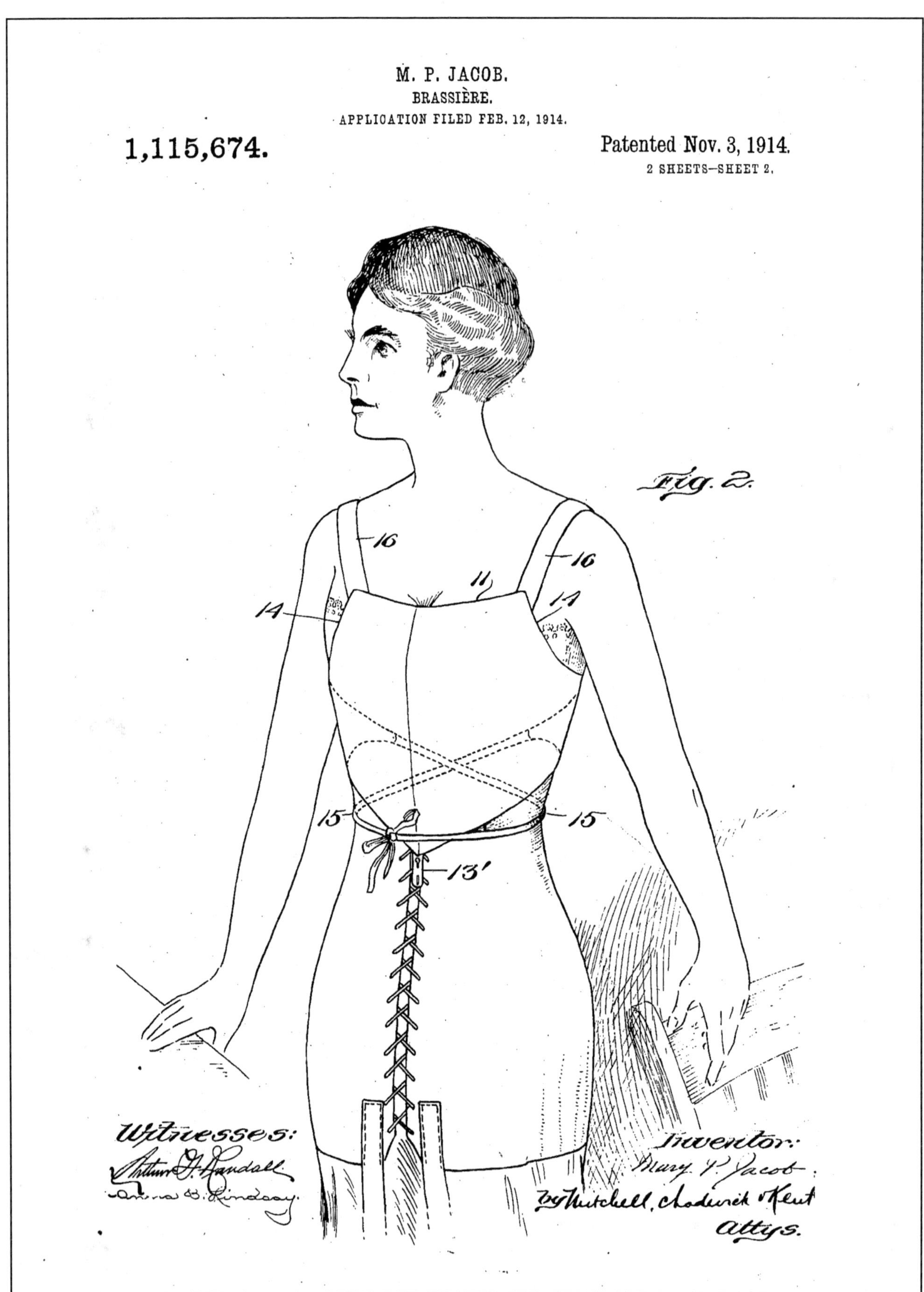

WOWidea! Corsets were bound to be updated! Oppressive clothing, long the bane of women during the 1800s, got a boost from this woman inventor. Mary Jacobs gained status with this improvement of the brassiere, or bra for short. This patent is acknowledged as the liberating design that only a woman could conceive or appreciate.

United States Patent [19]

Jacoby

[11] **3,938,115**

[45] **Feb. 11, 1976**

[54] **COMBINATION SMOKE AND HEAT DETECTOR ALARM**

[75] Inventor: **Sidney Jacoby,** Philadelphia, Pa.

[73] Assignee: **Evergard Fire Alarm Co., Inc.,** Philadelphia, Pa.

[22] Filed: **June 13, 1974**

[21] Appl. No.: **478,928**

[52] **U.S. Cl.** .... **340/237 S;** 116/106; 116/137 R

[51] **Int. Cl.**[2] **G08B 17/04;** G08B 17/10

[58] **Field of Search** .... 340/237 S, 227 R, 227.1, 340/228.5, 229, 404, 405, 406; 116/65, 70, 106, 114 Y, 142 FP, 2, 3, 5, 67 R, 101, 102, 103, 140, 137 R

[56] **References Cited**

UNITED STATES PATENTS

3,079,886	3/1963	Green, Jr.	116/106
3,109,409	11/1963	Demay	340/229 UX
3,119,368	1/1964	Barnard	340/229 UX
3,153,226	10/1964	Jensen	340/229 X
3,223,068	12/1965	Van Winkle	116/65

Primary Examiner—John W. Caldwell
Assistant Examiner—Daniel Myer
Attorney, Agent, or Firm—Weiser, Stapler & Spivak

[57] **ABSTRACT**

A combination smoke and heat detector alarm including a self-contained stored energy source in the form of a cylinder of compressed gas. A T-fitting connects to the cylinder and feeds separate conduit systems leading to individual sounding devices. A fusible element is interposed in one of the conduit systems to automatically permit transfer of the compressed gas to a first sounding device upon the presence of elevated temperatures. A solenoid operated switch is interposed in the other conduit system to normally prevent the flow of gas. The solenoid is responsive to a smoke detector and is wired to open the solenoid valve upon sensing the presence of a predetermined concentration of smoke.

12 Claims, 1 Drawing Figure

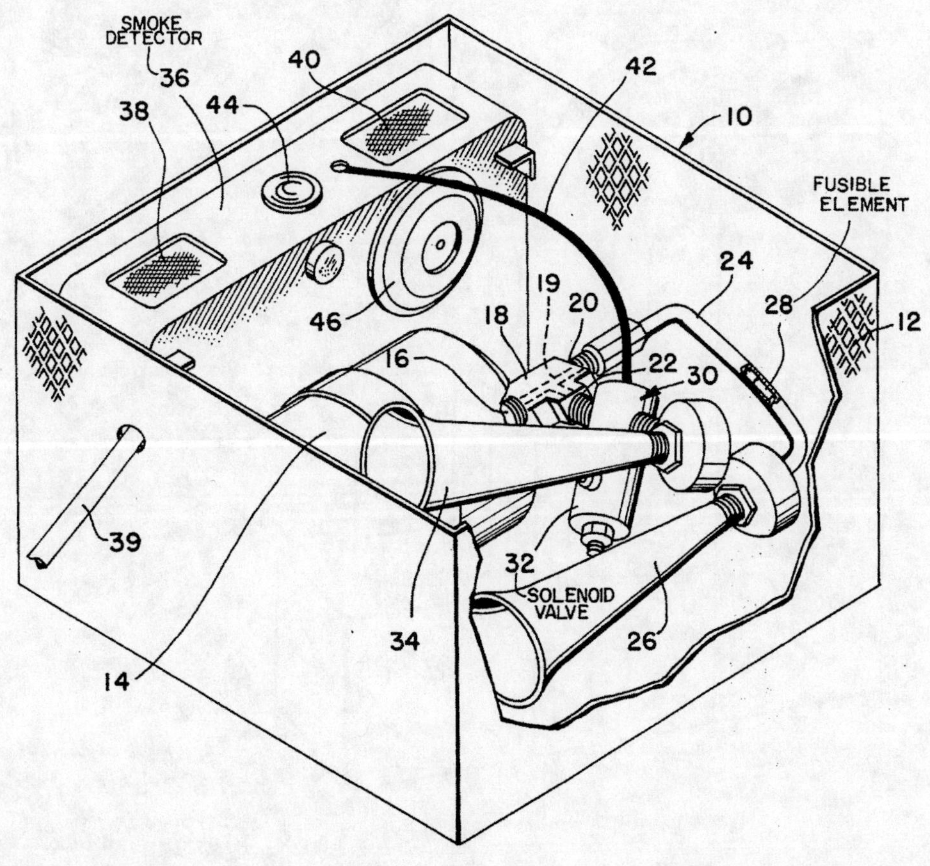

WOWidea! Early detection saves lives! This combination smoke alarm and heat detector invented by Sidney Jacoby was more than a metal box filled with compressed gas (an energy source), electronics, and horns. His innovative, self-powered, and lower cost design led to making basic life safety at home an affordable necessity.

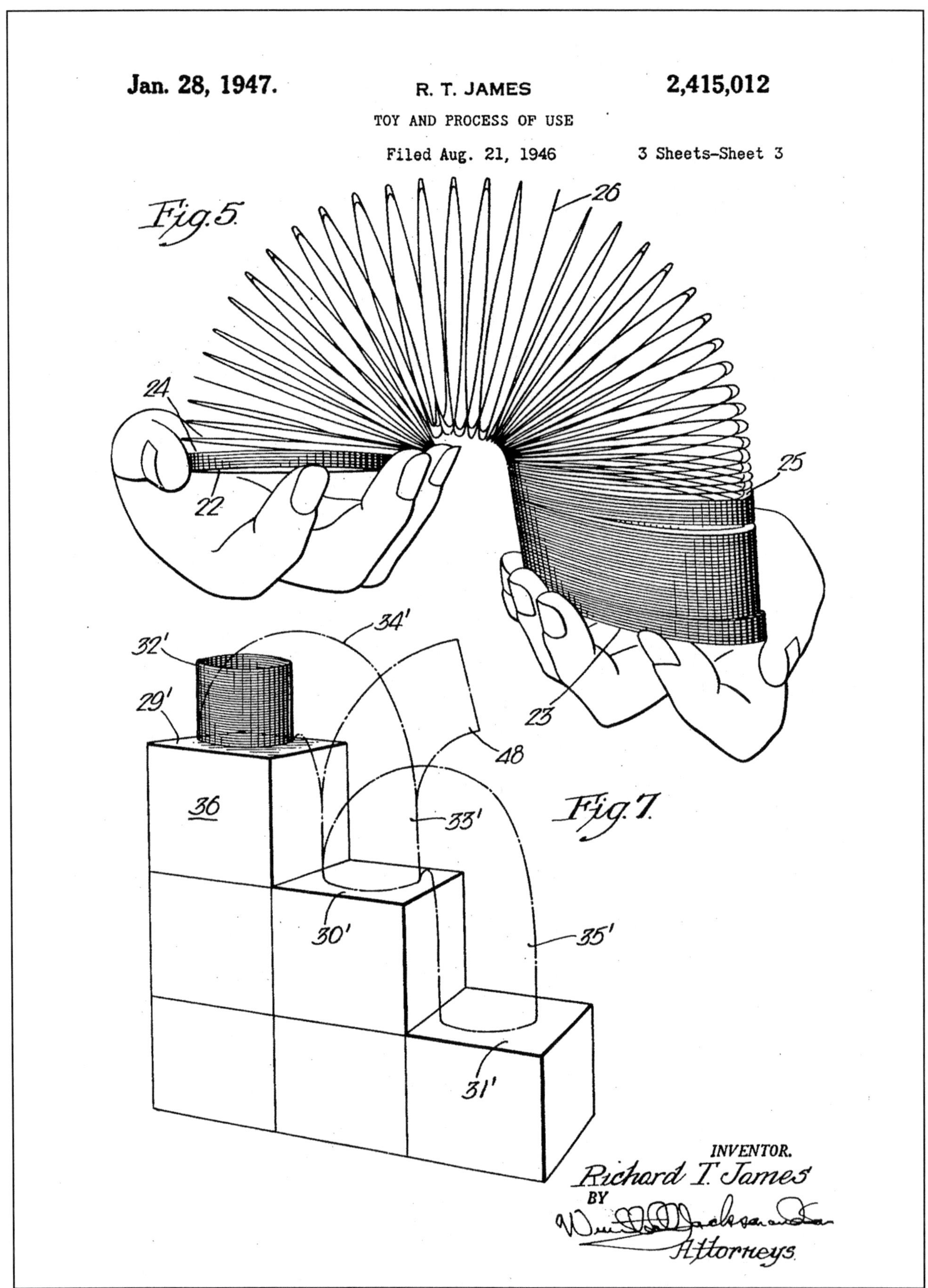

WOWidea! Toy springs to life by accident! Richard James, a Philadelphia Naval Shipyard worker, noticed that a torsion spring that accidentally fell to the floor acted "funny". Experiments took two years before Slinky® was demonstrated at a local department store. Since then, hundreds of millions of various styles have been sold.

United States Patent [19]

Jarvik

[11] **4,173,796**

[45] **Nov. 13, 1979**

[54] **TOTAL ARTIFICIAL HEARTS AND CARDIAC ASSIST DEVICES POWERED AND CONTROLLED BY REVERSIBLE ELECTROHYDRAULIC ENERGY CONVERTERS**

[75] Inventor: **Robert K. Jarvik,** Salt Lake City, Utah

[73] Assignee: **University of Utah,** Salt Lake City, Utah

[21] Appl. No.: **858,921**

[22] Filed: **Dec. 9, 1977**

[51] **Int. Cl.**[2] **A61F 1/24**

[52] **U.S. Cl.** .... **3/1.7**; 417/390; 417/389; 417/423 R

[58] **Field of Search** .... 3/1.7, 1; 128/1 D, DIG. 3; 417/390, 389, 394, 395, 423 R

[56] **References Cited**

U.S. PATENT DOCUMENTS

3,568,214	3/1971	Goldschmied	3/1.7
3,572,979	3/1971	Morton	3/1.7 X
3,633,217	1/1972	Lance	3/1.7
3,636,570	1/1972	Nielson	3/1.7
3,783,453	1/1974	Bolie	3/1.7
3,874,002	4/1975	Kurpanek	3/1.7

OTHER PUBLICATIONS

The Development of an Intrapericardial Cardiac Replacement Phase II, by W. H. Burns, et al., Transactions American Society for Artificial Internal Organs, vol. XII, 1966, pp. 272-274.

Primary Examiner—Ronald L. Frinks

[57] **ABSTRACT**

Total artificial hearts and circulatory assist devices, including left ventricular assist devices. The invention relates to electrohydraulic energy converter systems, whereby electric energy from a power source, which may be a battery or other source, is converted into hydraulic power capable of actuating diaphragm, sack, axi-symmetric or other types of blood pumps.

11 Claims, 12 Drawing Figures

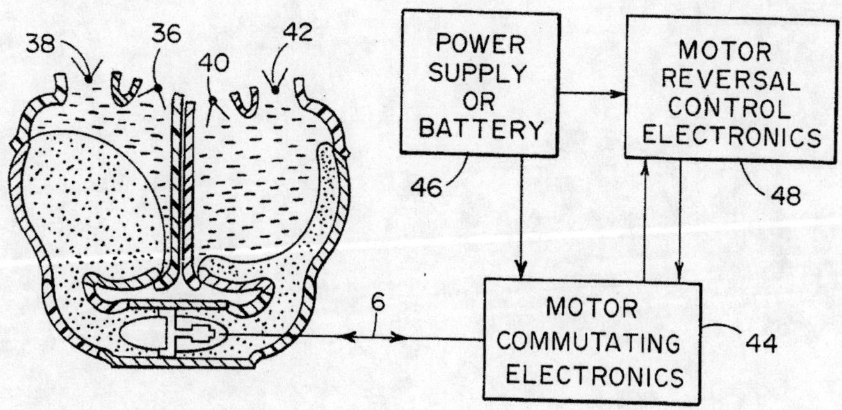

WOWidea! Keep trying! The lure of sustaining life by creating an artificial heart pump had consumed the efforts of over 100 American inventors and countless others but none were ever formally recognized with U.S. patents. It was medical device pioneer, Dr. Robert Jarvik, who put it all together to design a series of practical solutions.

(No Model.)

W. L. JUDSON.

CLASP LOCKER OR UNLOCKER FOR SHOES.

No. 504,038. Patented Aug. 29, 1893.

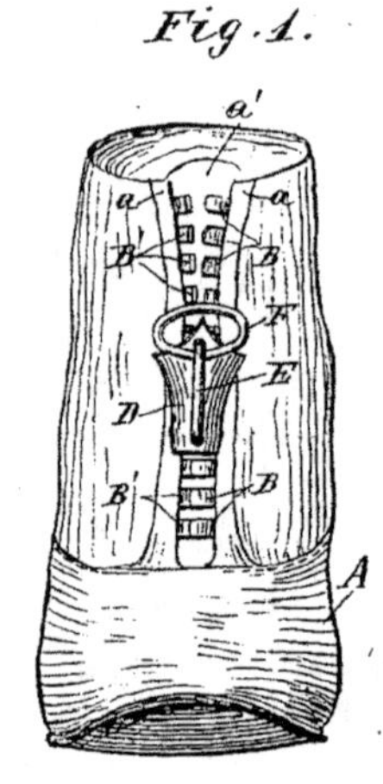

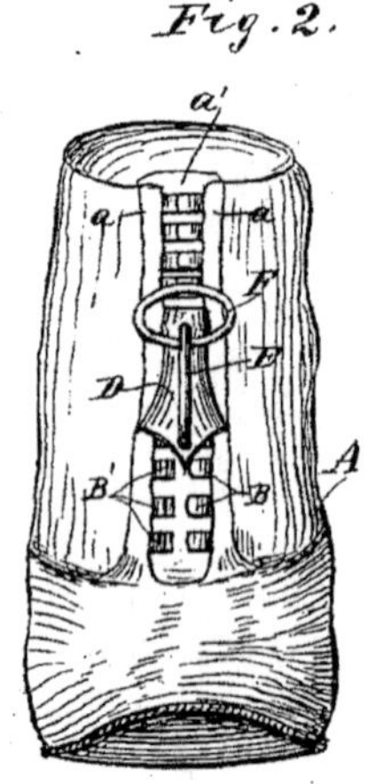

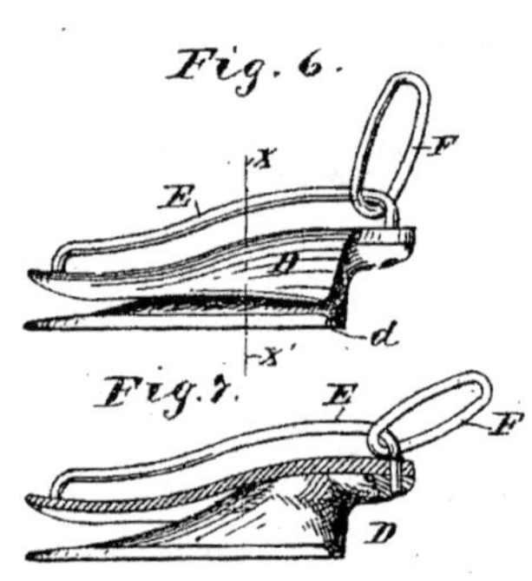

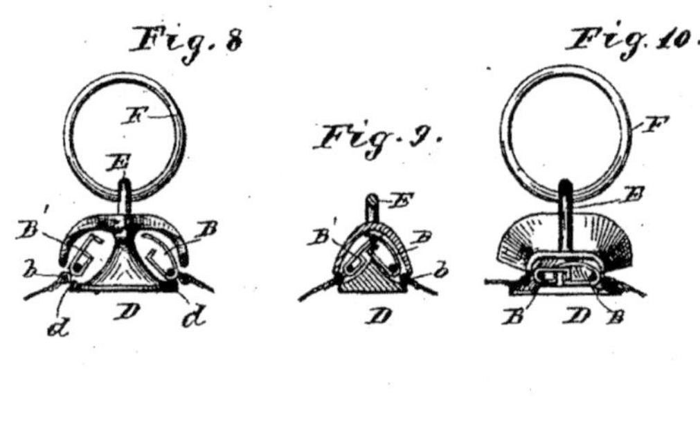

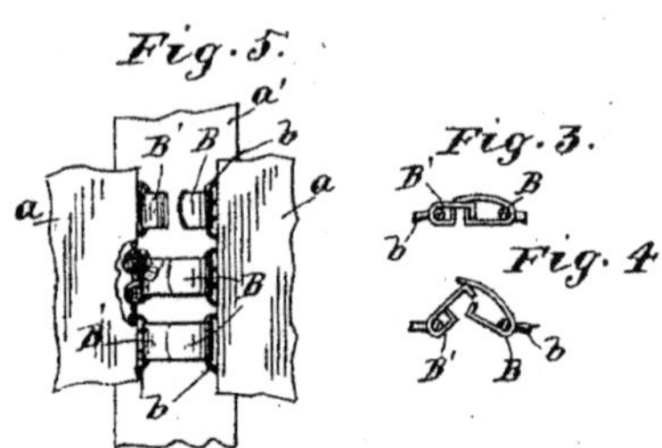

Witnesses.
A. U. Opsahl.
E. F. Elmore

Inventor.
Whitcomb L. Judson
By his Attorney.
Jas. F. Williamson

WOWidea! Continual improvement unlocks the zipper! Whitcomb Judson, a street railway innovator, turned his attention to mechanical fasteners, only to have the clothing industry reject his ideas for over 40 years. It was left to an employee, G. Sundback, to make them practical for use in rubber boots; Closing them made the sound of "zip".

No. 748,284. PATENTED DEC. 29, 1903.

J. KARWOWSKI.

METHOD OF PRESERVING THE DEAD.

APPLICATION FILED OCT. 13, 1903.

NO MODEL.

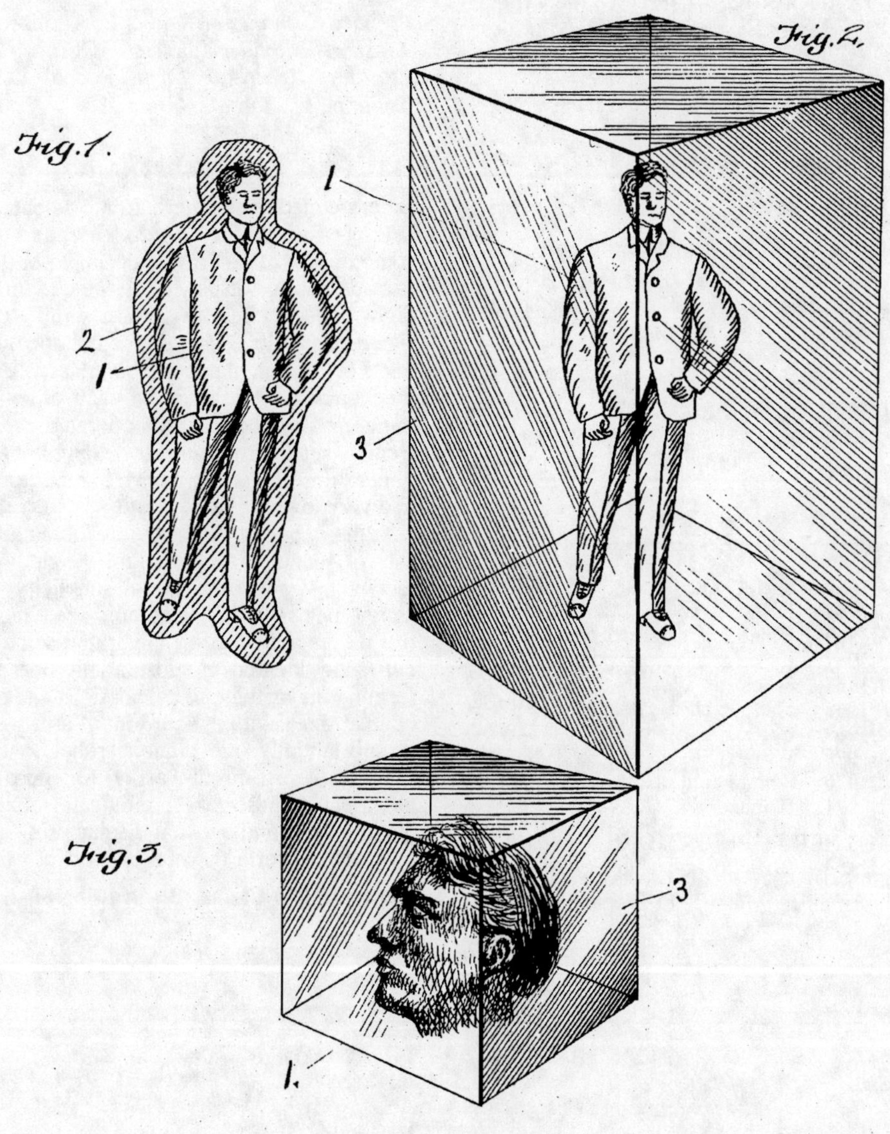

Inventor

Joseph Karwowski.

Witnesses

R. A. Boswell

A. L. Hough

By Franklin H. Hough

Attorney

WOWidea! The ultimate paper weight! Was Joe Karwowski a-head of his time with this intriguing, and some would say disgusting, method of preserving the "life-like" condition of a person or part thereof? He envisioned first surrounding the corpse with a layer of silica, then encasing the remains in glass. Kids, do not try this at home!

United States Patent [19] [11] **3,819,921**

Kilby et al. [45] **June 25, 1974**

[54] **MINIATURE ELECTRONIC CALCULATOR**

[75] Inventors: **Jack S. Kilby; Jerry D. Merryman; James H. Van Tassel,** all of Dallas, Tex.

[73] Assignee: **Texas Instruments Incorporated,** Dallas, Tex.

[22] Filed: **Dec. 21, 1972**

[21] Appl. No.: **317,493**

Related U.S. Application Data

[63] Continuation of Ser. No. 143,192, May 13, 1971, abandoned, which is a continuation of Ser. No. 671,777, Sept. 29, 1967, abandoned.

[52] **U.S. Cl.** **235/156,** 235/159
[51] **Int. Cl.** **G06f 7/38**
[58] **Field of Search** .... 235/156, 159, 145; 340/365, 347 DD; 219/201

[56] **References Cited**

UNITED STATES PATENTS

2,932,816	4/1960	Stiefel et al.	340/347 DD
3,140,031	7/1964	Fitch	226/109
3,315,069	4/1967	Bohm	235/156 X
3,331,954	7/1967	Kinzie et al.	235/156
3,354,817	11/1967	Sakurai et al.	101/93
3,405,392	10/1968	Milne et al.	235/156 X
3,430,226	2/1969	Chow	340/347
3,509,329	4/1970	Wang et al.	235/156
3,548,179	12/1970	Kimura et al.	235/156
3,553,445	1/1971	Hernandez	235/156

OTHER PUBLICATIONS

National Technical Report, Vol. 12, No. 2, 1966, pp. 129–137, (Published by Matsushita Electric Industrial Co., Ltd., Japan).

Primary Examiner—Charles E. Atkinson
Assistant Examiner—James F. Gottman
Attorney, Agent, or Firm—Harold Levine; Edward Connors, Jr.; John G. Graham

[57] **ABSTRACT**

Binary-coded decimal electronic calculator capable of adding, subtracting, multiplying and dividing with some degree of automatic decimal point placement to provide a visual display of answers of up to 12 decimal digits. The decimal digits are serially displayed at a speed compatible with the calculator operations. The parts of the calculator are so adapted electrically and mechanically in relation to each other to result in a minature portable battery operated calculator of extremely small dimensions for example the outside case dimensions of 4¼ inches by 6⅛ inches by 1¾ inches and very low weight of about 45 ounces, having a calculating capability only before obtainable in calculators of much larger size and weight while retaining mechanical and operational simplicity. Some significant aspects of the calculator are the primary electronics embodied in an integrated semiconductor circuit array located in substantially one plane for performing the arithmetic calculations and generating the control signals, a keyboard input arrangement located in substantially one plane parallel to the integrated semiconductor circuit array for producing unique electrical signals corresponding to number and command entries and a visual display using a semiconductor array, as for a thermal printer for printout.

59 Claims, 36 Drawing Figures

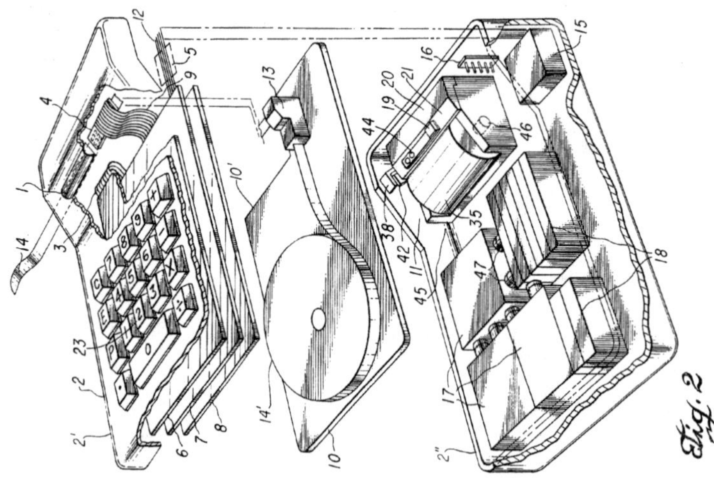

WOWidea! Hold it in your hand! Jack Kilby, Jerry Merriman and James Van Tassel of Texas Instruments were challenged to develop a breakthrough product with the latest technology, tiny integrated circuits. Of all the ideas they thought through, a hand-held calculator seemed like an ideal choice. They even stuffed a small printer inside.

Jan. 20, 1931. E. M. KNABUSCH ET AL 1,789,337

RECLINING CHAIR

Filed Jan. 24, 1929 3 Sheets-Sheet 3

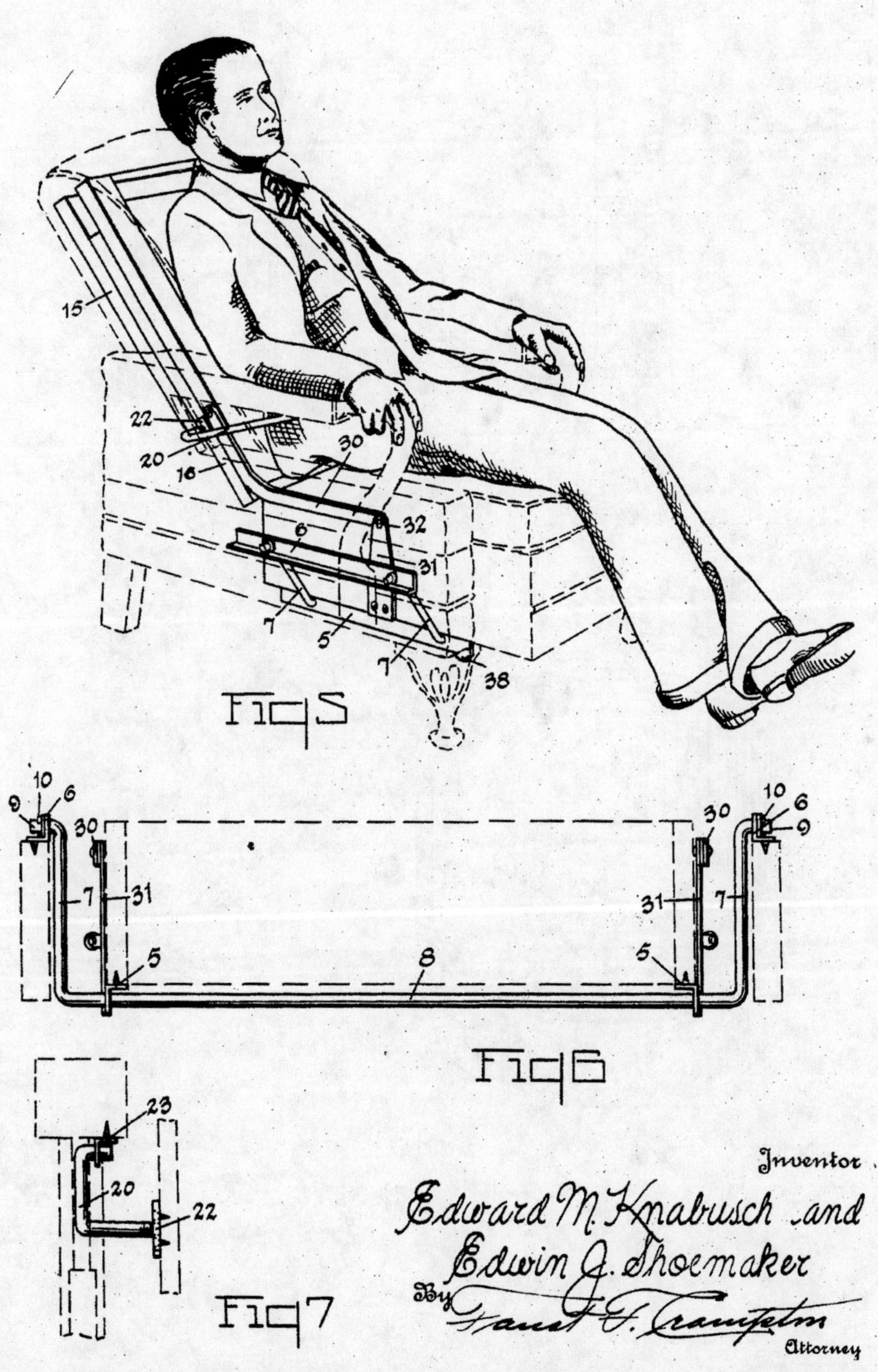

WOWidea! Being told “no” was a blessing in disguise. Cousins Edward Knabusch and Edwin Shoemaker were young furniture-making entrepreneurs from Monroe, MI, when they approached a department store buyer with their wooden-slat reclining porch chair. He wouldn’t buy it unless they upholstered it– the La-Z-Boy® was born!

June 9, 1953 J. H. KRAFT 2,641,545

MANUFACTURE OF SOFT SURFACE CURED CHEESE

Filed April 11, 1951

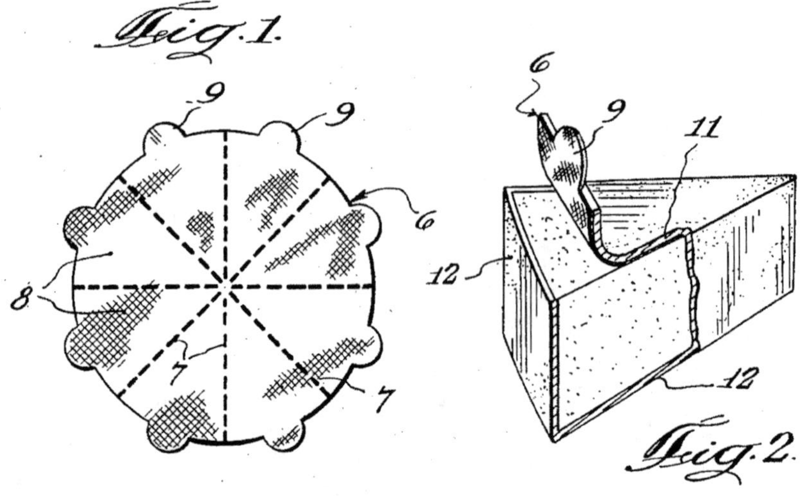

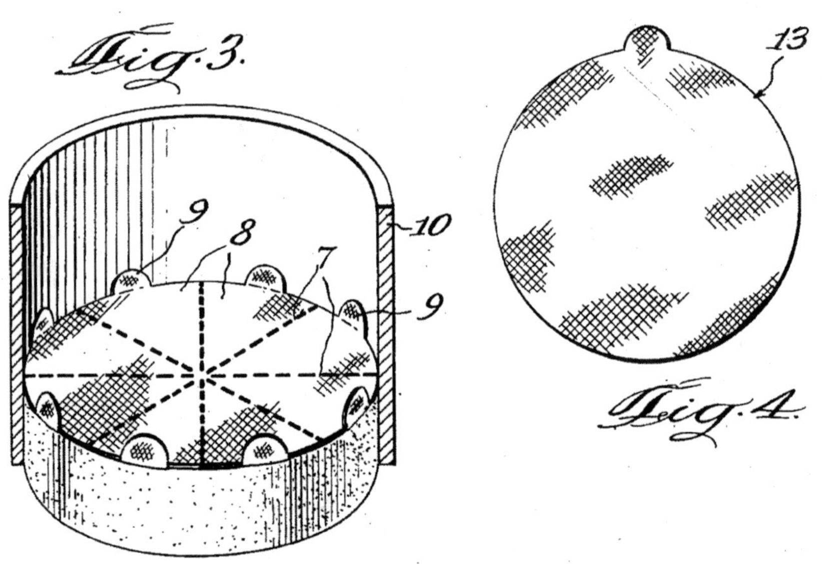

Inventor
John H. Kraft
By Soans Glaister & Anderson
Attorneys

WOWidea! Holiday baskets and lunch boxes wouldn't be complete without these little foil wrapped delights! John Kraft, whose brother James in 1916 patented the first "process cheese", was following in the family tradition of innovation. These food pioneers also gave us CHEEZ WHIZ® and the first individually wrapped slices of cheese.

(No Model.) 6 Sheets—Sheet 1.

S. LAKE.

SUBMARINE VESSEL.

No. 581,213. Patented Apr. 20, 1897.

Fig. 1.

Witnesses.
Dennis Sumby.
Robert Everett.

Inventor.
Simon Lake.
By James L. Norris.
Atty.

WOWidea! Walk on the bottom of the ocean! From the mid-1600s through the 1800s, submarines of various designs had been used to fight naval battles. However, it was left to Simon Lake and a competitor, John Holland, to figure out the most practical method of getting to the bottom of fishy things… and living to tell the tale.

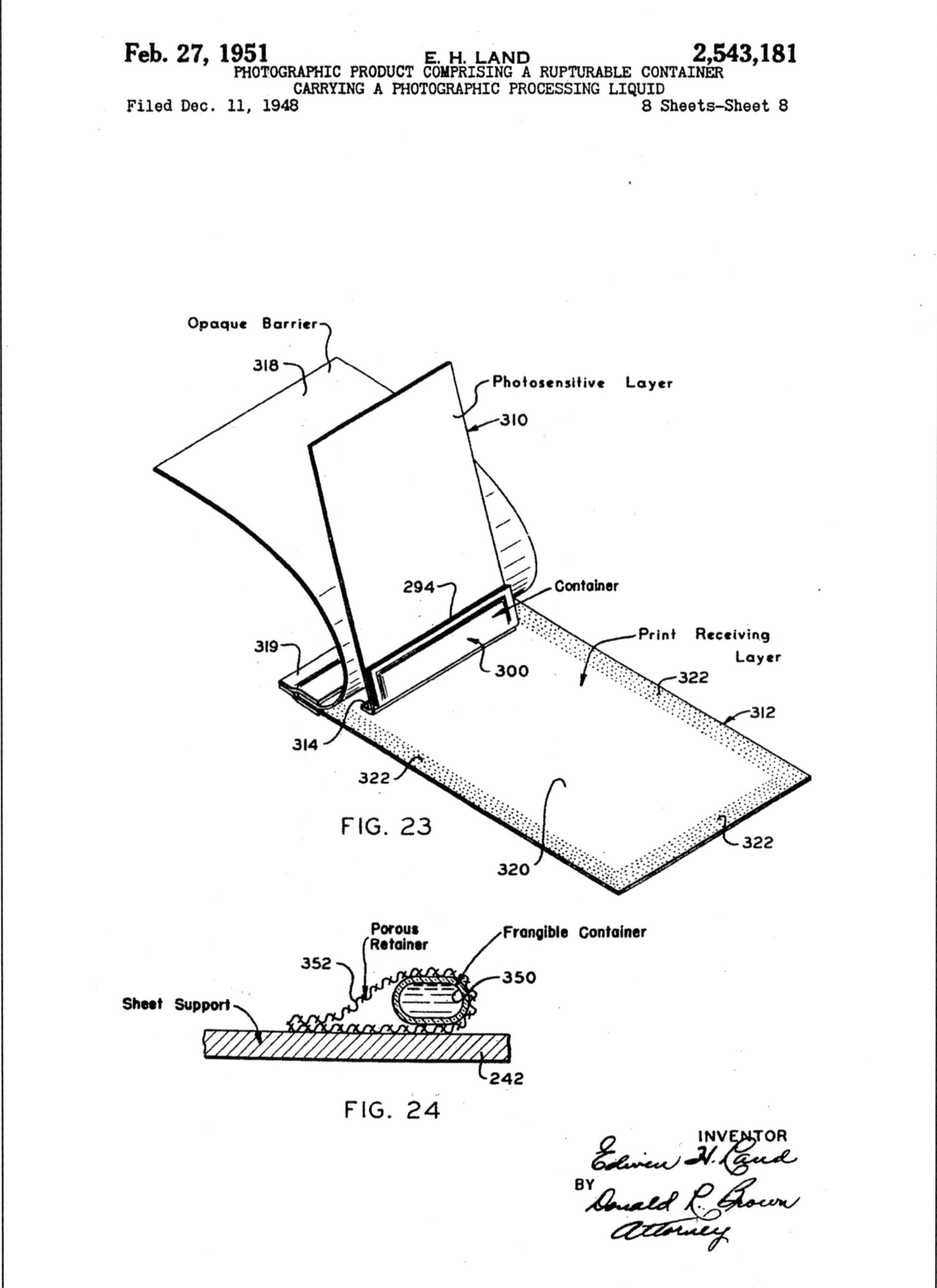

WOWidea! One step, one minute, one photograph! A pioneer in the polarization of light, Edwin Land co-founded what became the Polaroid Corporation. After exposure to light, a chemical-laden negative was pressed against a positive to produce an "instant" photo. The first version was sepia, then black and white, and ultimately full color.

Feb. 20, 1934. E. O. LAWRENCE **1,948,384**

METHOD AND APPARATUS FOR THE ACCELERATION OF IONS

Filed Jan. 26, 1932 2 Sheets-Sheet 1

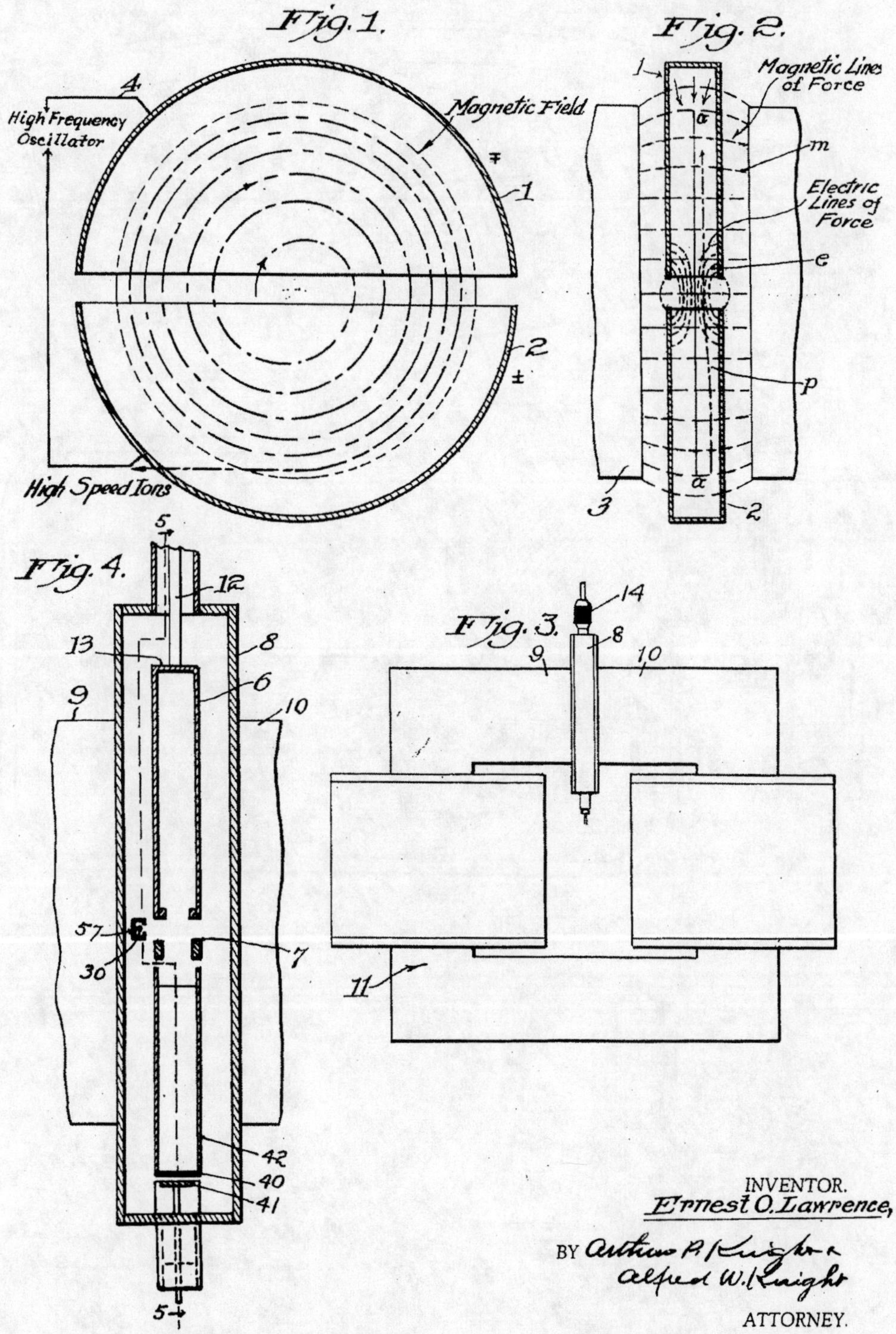

INVENTOR.
Ernest O. Lawrence,

BY Arthur P. Knight
Alfred W. Knight

ATTORNEY.

WOWidea! Create a high-speed raceway that shoots and separates atoms! Born in South Dakota, Ernest Lawrence eventually found his way to the University of California as a faculty member. Here he developed what's now known as the "Cyclotron", which ultimately produced enough radioactive uranium for the first nuclear reactor.

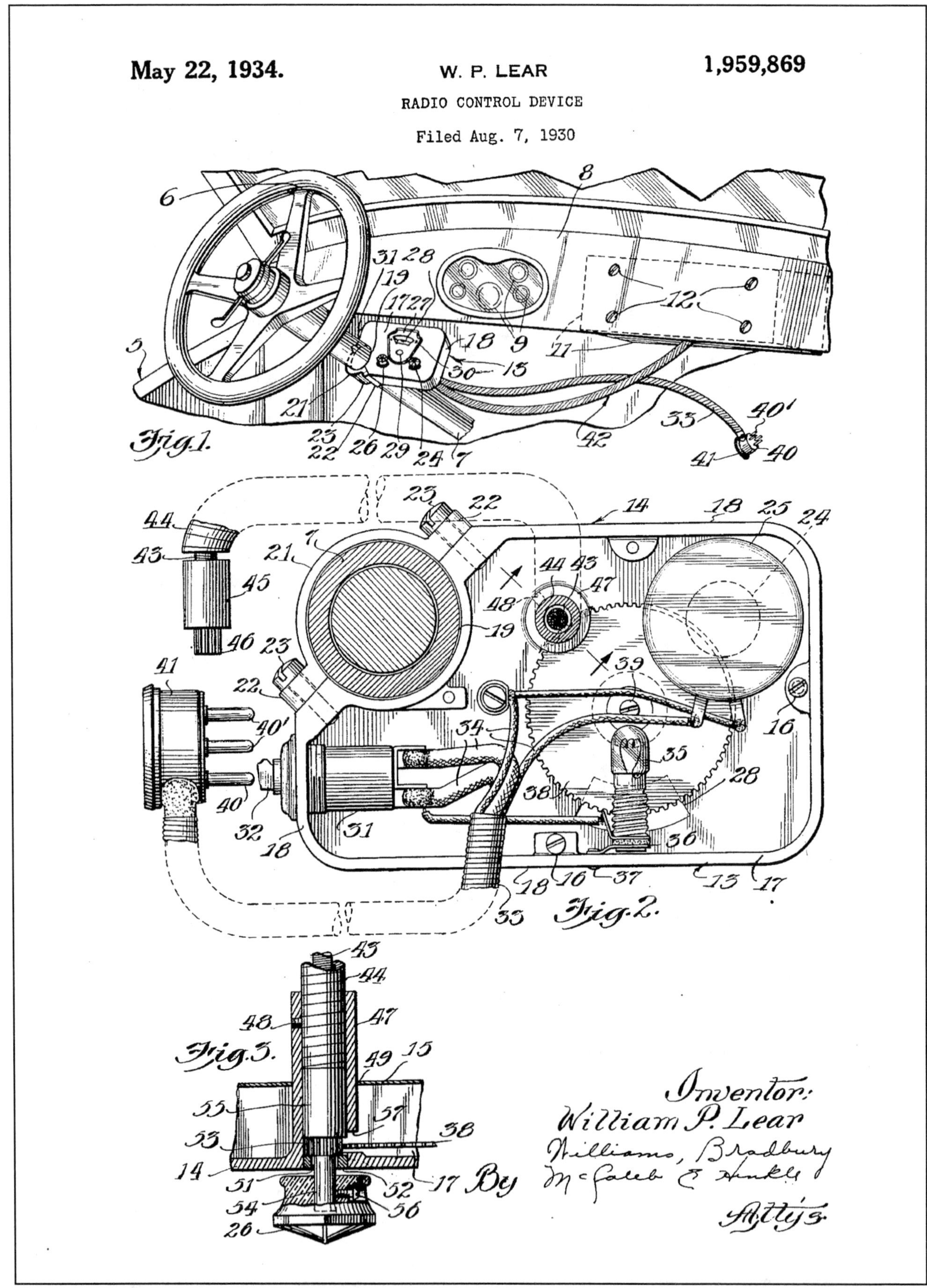

WOWidea! Driving music, what a thought! William Lear devised the first static-free radio that could be put into a car with a running motor. He quickly licensed the patent to a firm which would go on to become an electronics powerhouse, Motorola®. He's also widely known for the Learjet® and lesser known for his 1970's 8-track tape.

United States Patent [19]

Leder et al.

[11] **Patent Number: 4,736,866**

[45] **Date of Patent: Apr. 12, 1988**

[54] **TRANSGENIC NON-HUMAN MAMMALS**

[75] Inventors: **Philip Leder,** Chestnut Hill, Mass.; **Timothy A. Stewart,** San Francisco, Calif.

[73] Assignee: **President and Fellows of Harvard College,** Cambridge, Mass.

[21] Appl. No.: **623,774**

[22] Filed: **Jun. 22, 1984**

[51] **Int. Cl.**[4] **C12N 1/00;** C12Q 1/68; C12N 15/00; C12N 5/00

[52] **U.S. Cl.** **800/1;** 435/6; 435/172.3; 435/240.1; 435/240.2; 435/320; 435/317.1; 935/32; 935/59; 935/70; 935/76; 935/111

[58] **Field of Search** 435/6, 172.3, 240, 317, 435/320, 240.1, 240.2; 935/70, 76, 59, 111, 32; 800/1

[56] **References Cited**

U.S. PATENT DOCUMENTS

4,535,058 8/1985 Weinberg et al. 435/91
4,579,821 4/1986 Palmiter et al. 435/240

OTHER PUBLICATIONS

Ucker et al, Cell 27:257–266, Dec. 1981.
Ellis et al, Nature 292:506–511, Aug. 1981.
Goldfarb et al, Nature 296:404–409, Apr. 1981.
Huang et al, Cell 27:245–255, Dec. 1981.
Blair et al, Science 212:941–943, 1981.
Der et al, Proc. Natl. Acad. Sci. USA 79:3637–3640, Jun. 1982.
Shih et al, Cell 29:161–169, 1982.
Gorman et al, Proc. Natl. Acad. Sci. USA 79:6777–6781, Nov. 1982.
Schwab et al, EPA–600/9–82–013, Sym: Carcinogen, Polynucl. Aromat. Hydrocarbons Mar. Environ., 212–32 (1982).
Wagner et al. (1981) Proc. Natl. Acad. Sci USA 78, 5016–5020.
Stewart et al. (1982) Science 217, 1046–8.
Costantini et al. (1981) Nature 294, 92–94.
Lacy et al. (1983) Cell 34, 343–358.
McKnight et al. (1983) Cell 34, 335.
Binster et al. (1983) Nature 306, 332–336.
Palmiter et al. (1982) Nature 300, 611–615.
Palmiter et al. (1983) Science 222, 814.
Palmiter et al. (1982) Cell 29, 701–710.

Primary Examiner—Alvin E. Tanenholtz
Attorney, Agent, or Firm—Paul T. Clark

[57] **ABSTRACT**

A transgenic non-human eukaryotic animal whose germ cells and somatic cells contain an activated oncogene sequence introduced into the animal, or an ancestor of the animal, at an embryonic stage.

12 Claims, 2 Drawing Sheets

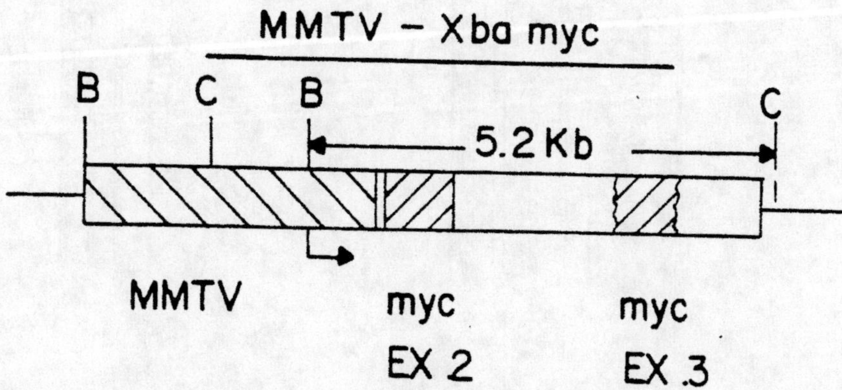

WOWidea! A mouse could roar! Philip Leder and Timothy A. Stewart, using genetic engineering, created the first mammals to be patented. Religious and political groups soon sounded alarm bells and the race to stop animal patenting was on. Interestingly, these particular mice have become instrumental in finding cures for human cancers.

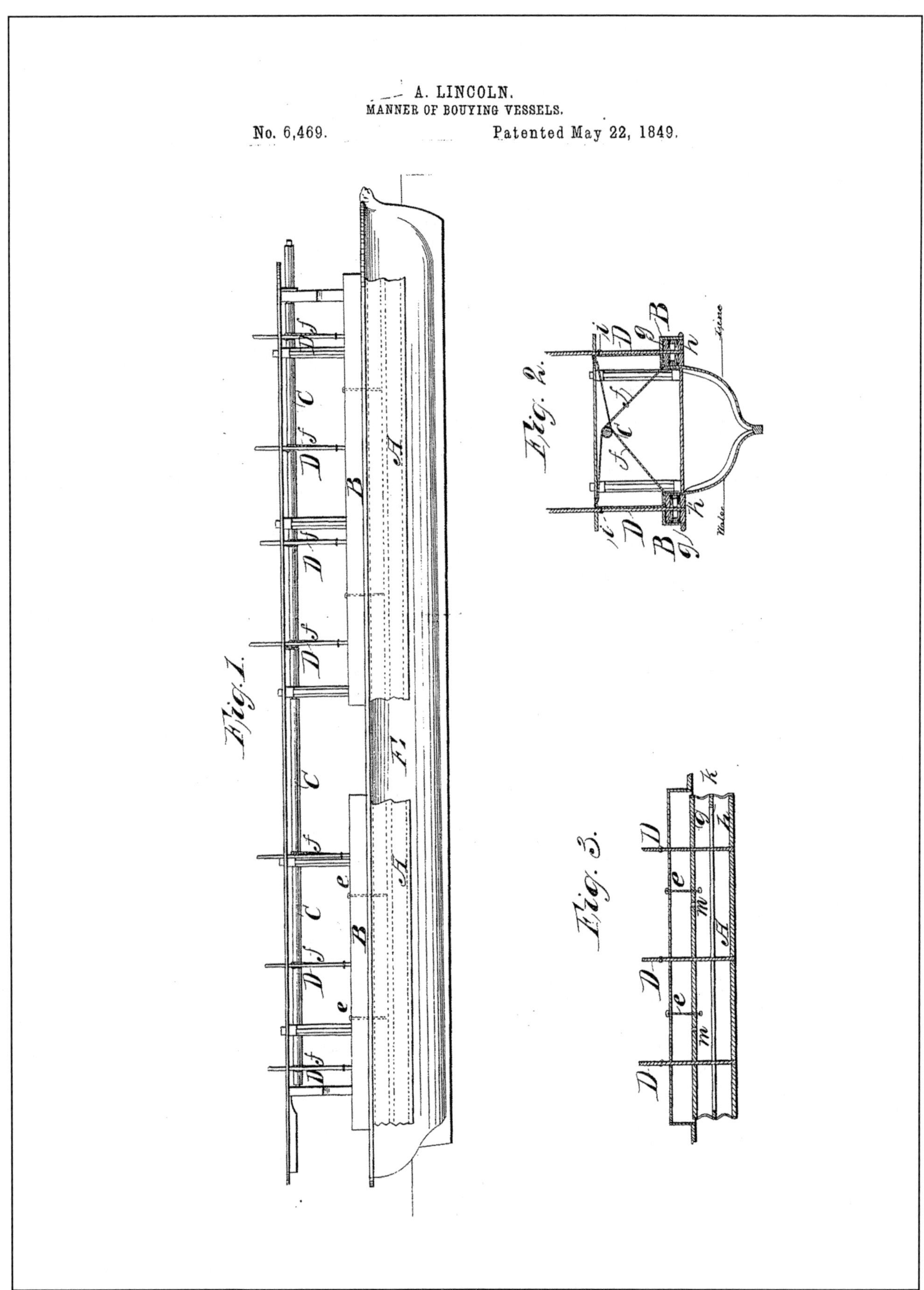

WOWidea! Future U.S. President applies his boyhood ingenuity while working on the Mississippi River! Noticing that boats often got caught on the shoals hidden beneath the river, Abraham Lincoln conceived this system of buoys. Later in life, as a leader, his ability to seek solutions to other vexing issues would become his legacy.

(No Model.)

J. J. LOUD.

PEN.

No. 392,046. Patented Oct. 30, 1888.

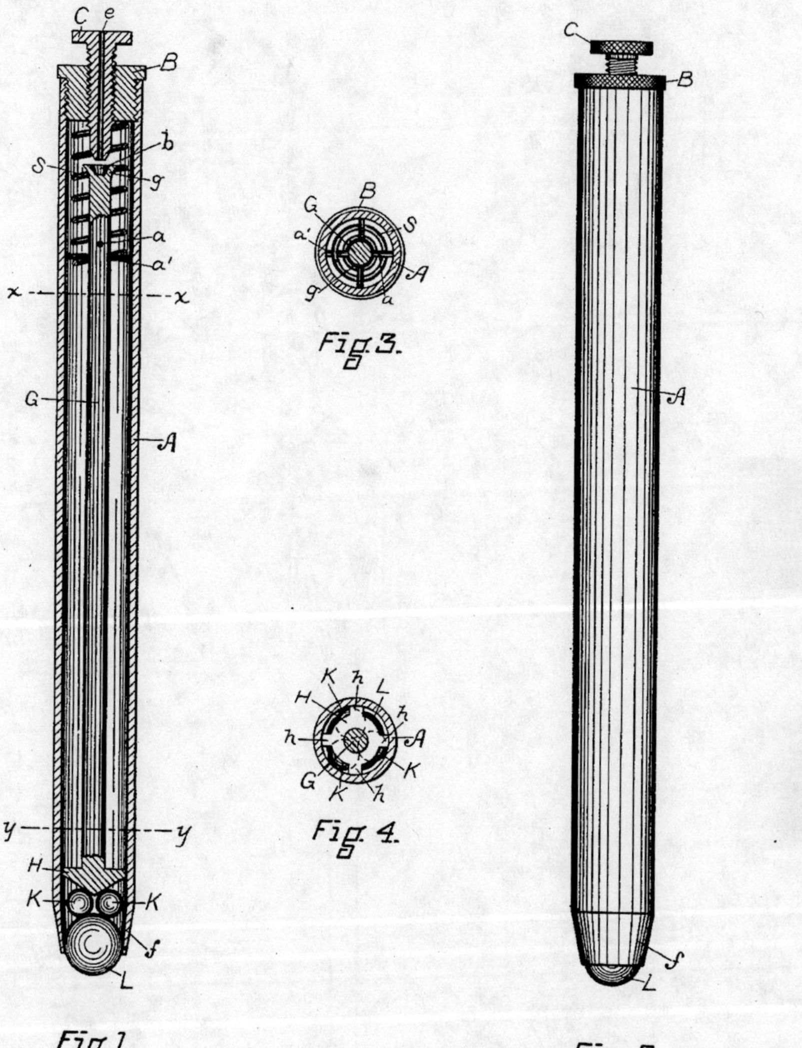

Fig. 1.

Fig. 2.

Fig. 3.

Fig. 4.

WITNESSES.

Albert E. Leach.

M. H. Thompson.

INVENTOR.

John J. Loud,

By his Attorney

Wm. B. H. Dowse.

WOWidea! Someone's got to be first! John Loud, an American leather tanner, patented this roller-ball-tip marking pen featuring a reservoir of ink and a roller ball that applied the thick ink to leather hides. His pen was never produced, nor were any of the other 350 patents for ball-type pens issued over the next thirty years.

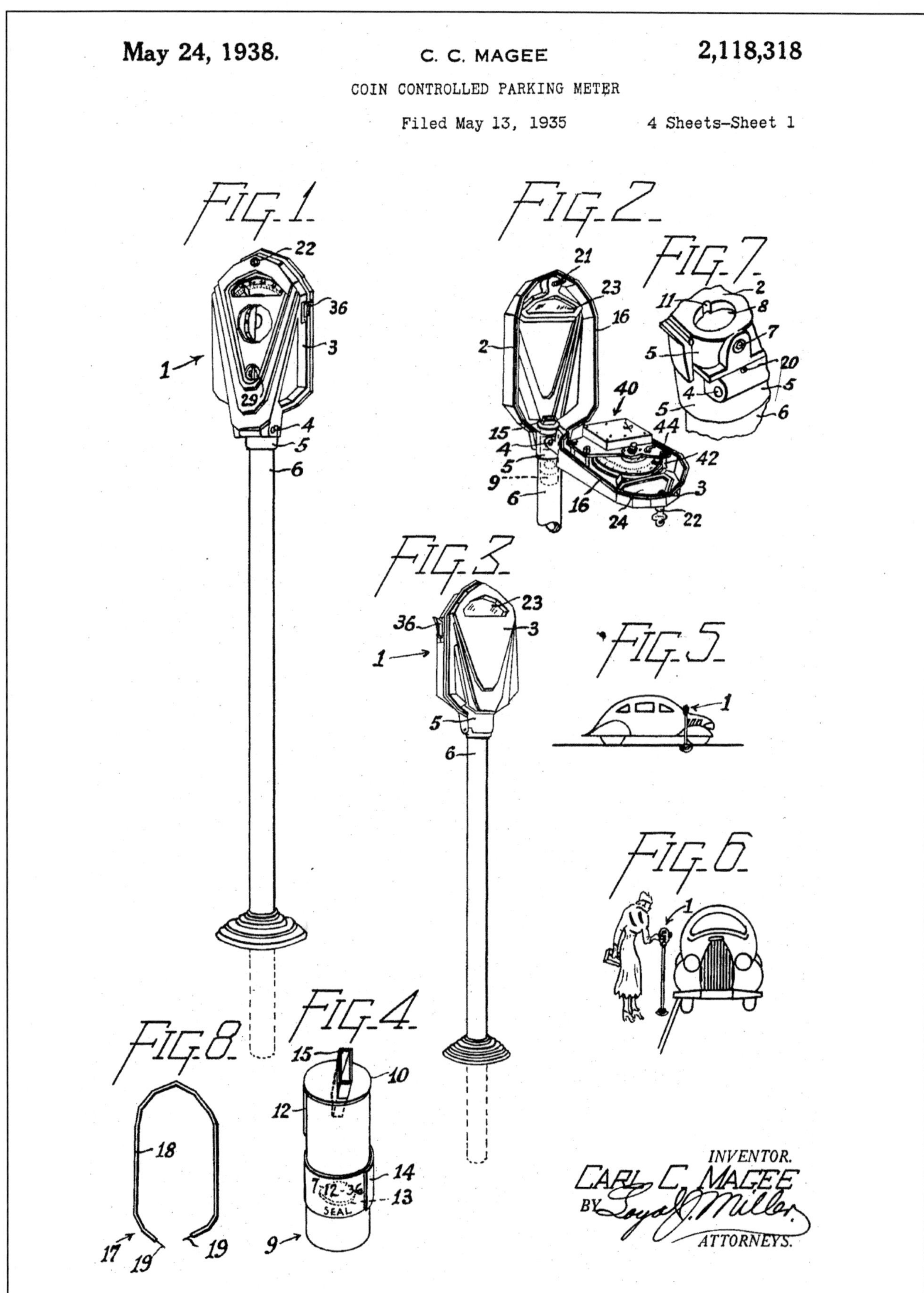

WOWidea! Small change can add up! Carl Magee, an Oklahoma City newspaper editor, was concerned about the lack of parking for downtown shoppers. Appointed to the Chamber of Commerce's committee to come up with a solution, he invented a pay-for-parking meter that revolutionized business districts around the globe.

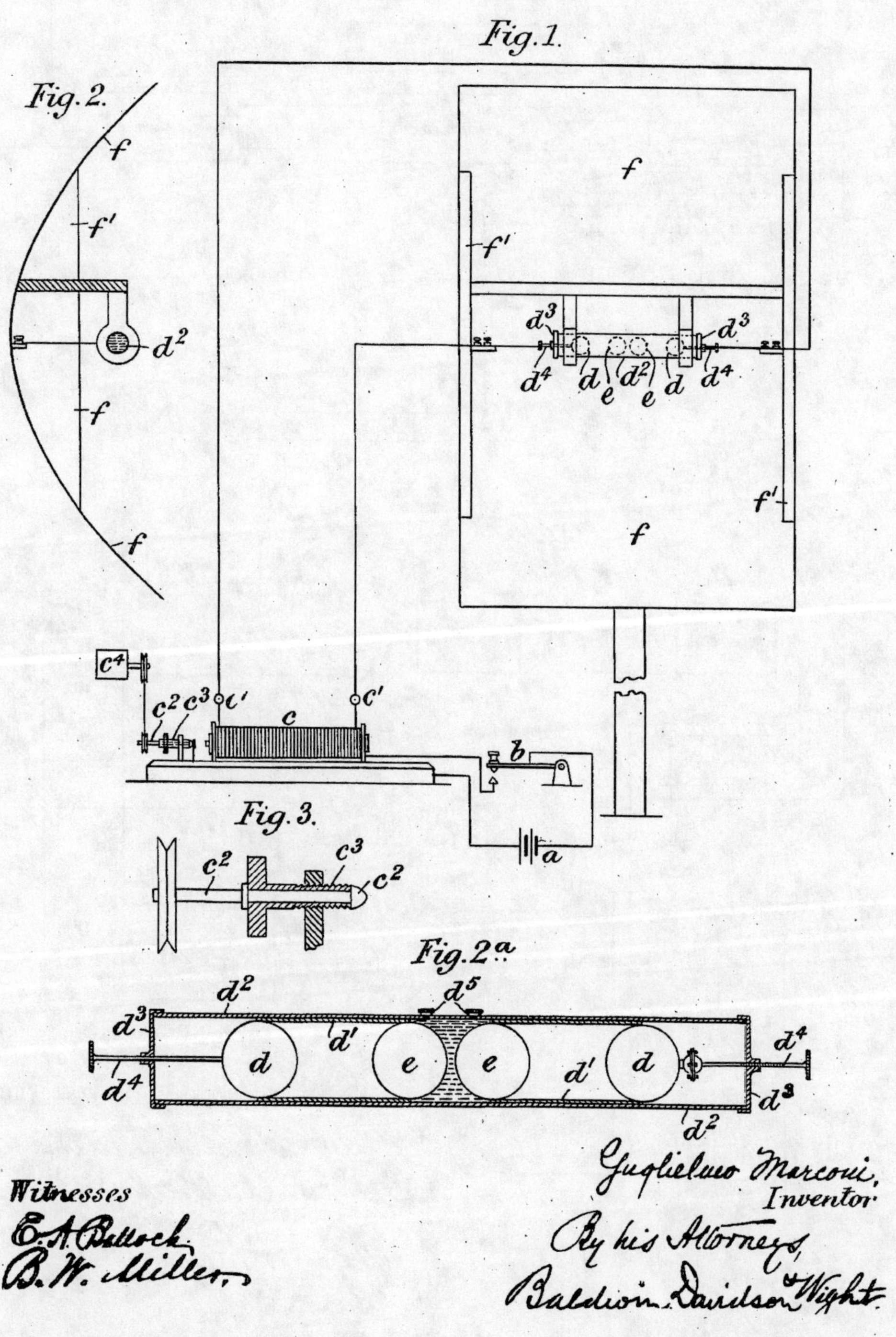

WOWidea! Look Ma, no wires! Guglielmo Marconi as a young adult experimented and built on the previous discoveries of others to make the wireless radio telegraph a reality 60 years after Morse had created the wired kind. Years later, to honor his Italian inventor-hero, another inventor named his successful toy cart the "*Radio* Flyer".

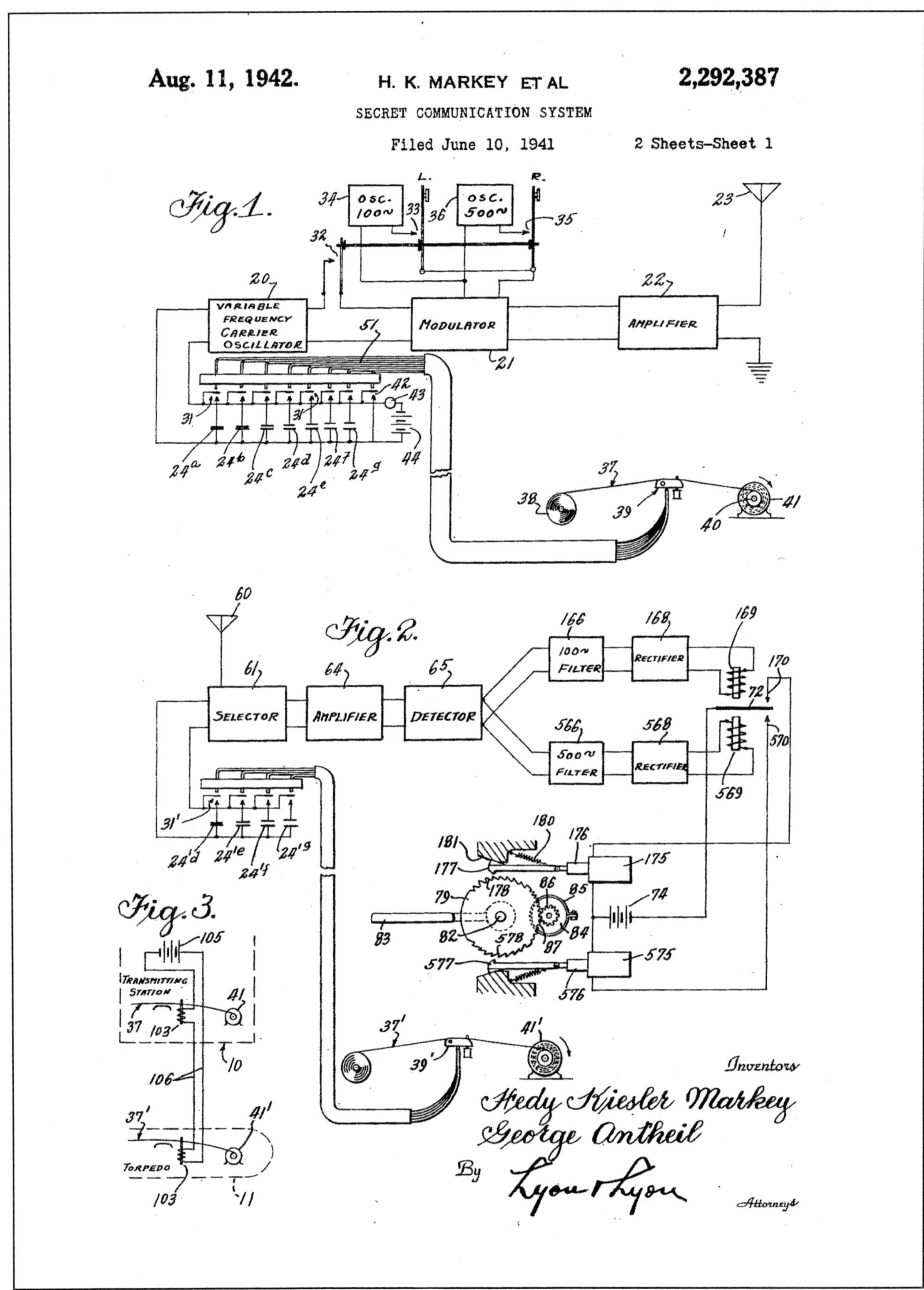

WOWidea! Leading lady thinks outside the box! Better known as actress Hedy Lamarr, this former wife of a World War II Nazi arms supplier, wanted to lend her learned expertise in warfare to the U.S. government. Along with her composer/writer co-inventor, they developed what would become an important form of wireless signaling.

J. L. Mason,

Glass Jar.

No 22,186. Patented Nov. 30, 1858.

Witnesses:

W. P. N. Fitzgerald

J. McMurray

Inventor:

John L. Mason

N. PETERS, PHOTO-LITHOGRAPHER, WASHINGTON, D. C.

WOWidea! The seal was the secret! Although hundreds of inventors obtained patents for fruit jars, probably the most well known was invented by John Landis Mason. He improved upon this shoulder-seal patent with an added rubber ring, but did not protect his name as well… it soon became synonymous with any competitor's canning jars.

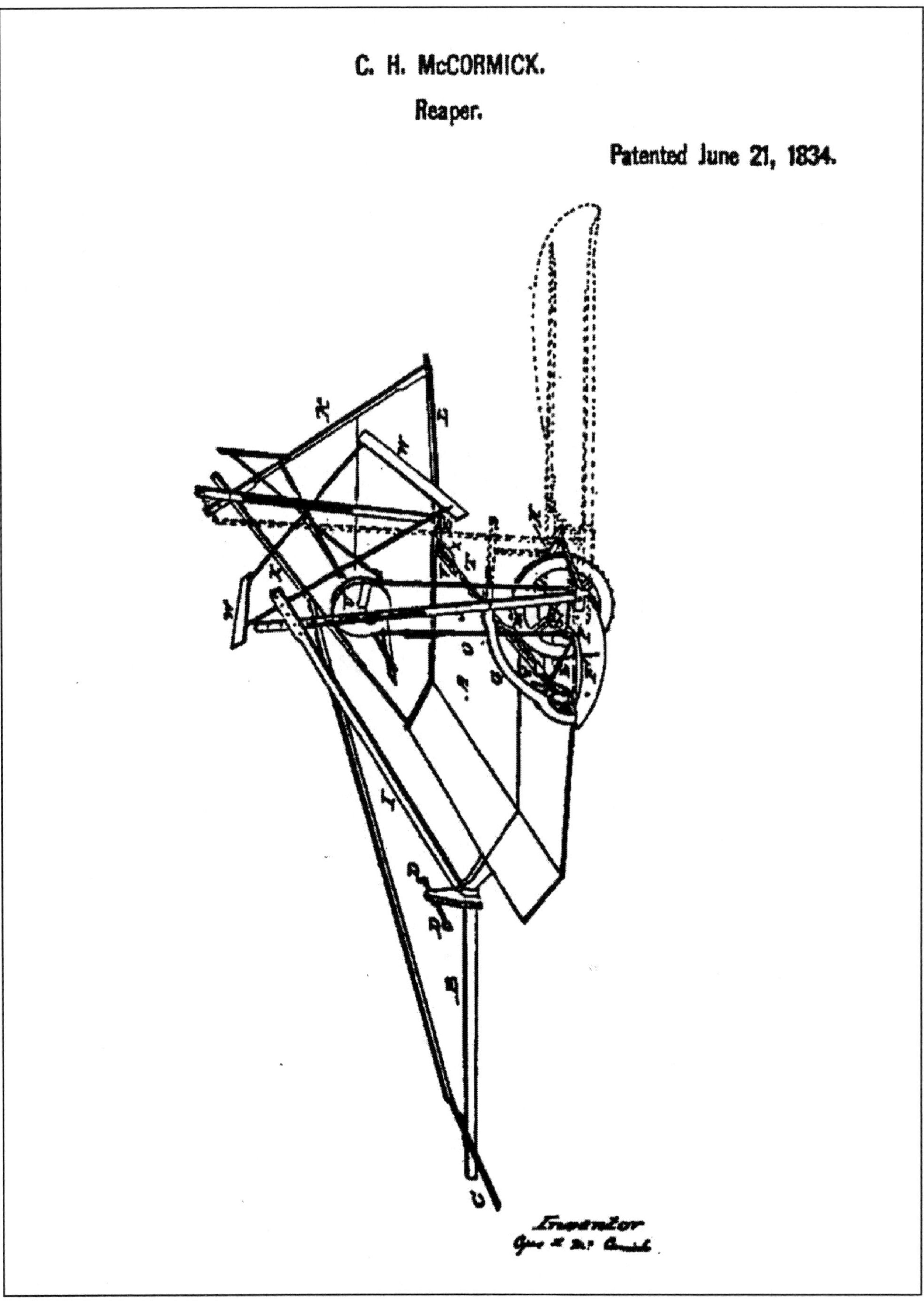

WOWidea! Let farmers reap the rewards! Cyrus McCormick perfected hay gathering and introduced a revolutionary form of business, which included standard parts, a guarantee, direct-mail advertising and testimonials. With eventual competition from Nantucket-based rival Obed Hussey, the American system of free enterprise boomed.

E. McCOY.

Lubricators.

No. 139,407. Patented May 27, 1873.

fig.1.

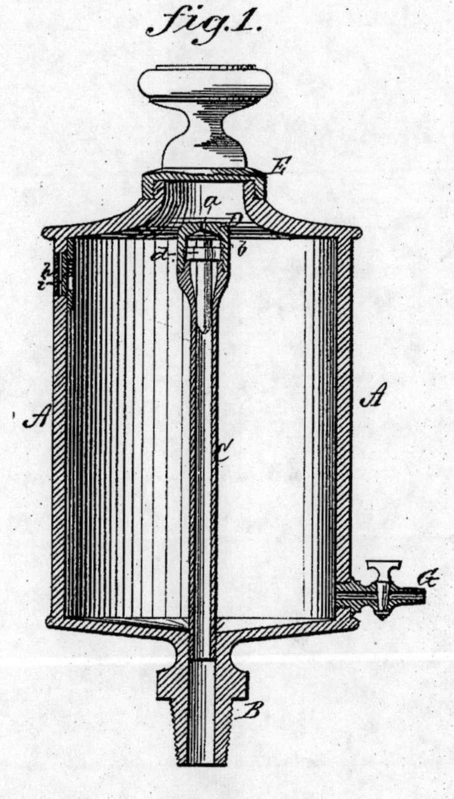

fig.2.

Witnesses
John A. Ellis
C. Alexander

Inventor
Elijah McCoy
Per
J. H. Alexander & Co
Attys

WOWidea! Parents' encouragement makes for success! Elijah McCoy's family moved to Canada to escape slavery and as a boy he loved to take things apart and put them back together. His first of 57 patents was for a very reliable lubricator that dripped oil onto train car couplings. Copied by competitors, buyers would ask for "the real McCoy".

DESIGN.

No. 37,431. PATENTED MAY 16, 1905.

F. McINTYRE.

LEAD PENCIL.

APPLICATION FILED NOV. 3, 1904.

WITNESSES:

INVENTOR

BY

his Attorney

WOWidea! The prototypical pencil! What a wonderfully simple and elegant statement this pencil design makes. Frank McIntyre is not the inventor of the lead pencil, that anonymous distinction goes back hundreds of years. Interestingly, writer Henry David Thoreau and his family were leading American pencil makers in the early 1800s.

(No Model.) 4 Sheets—Sheet 1.

O. MERGENTHALER.

MATRIX MAKING MACHINE.

No. 304,272. Patented Aug. 26, 1884.

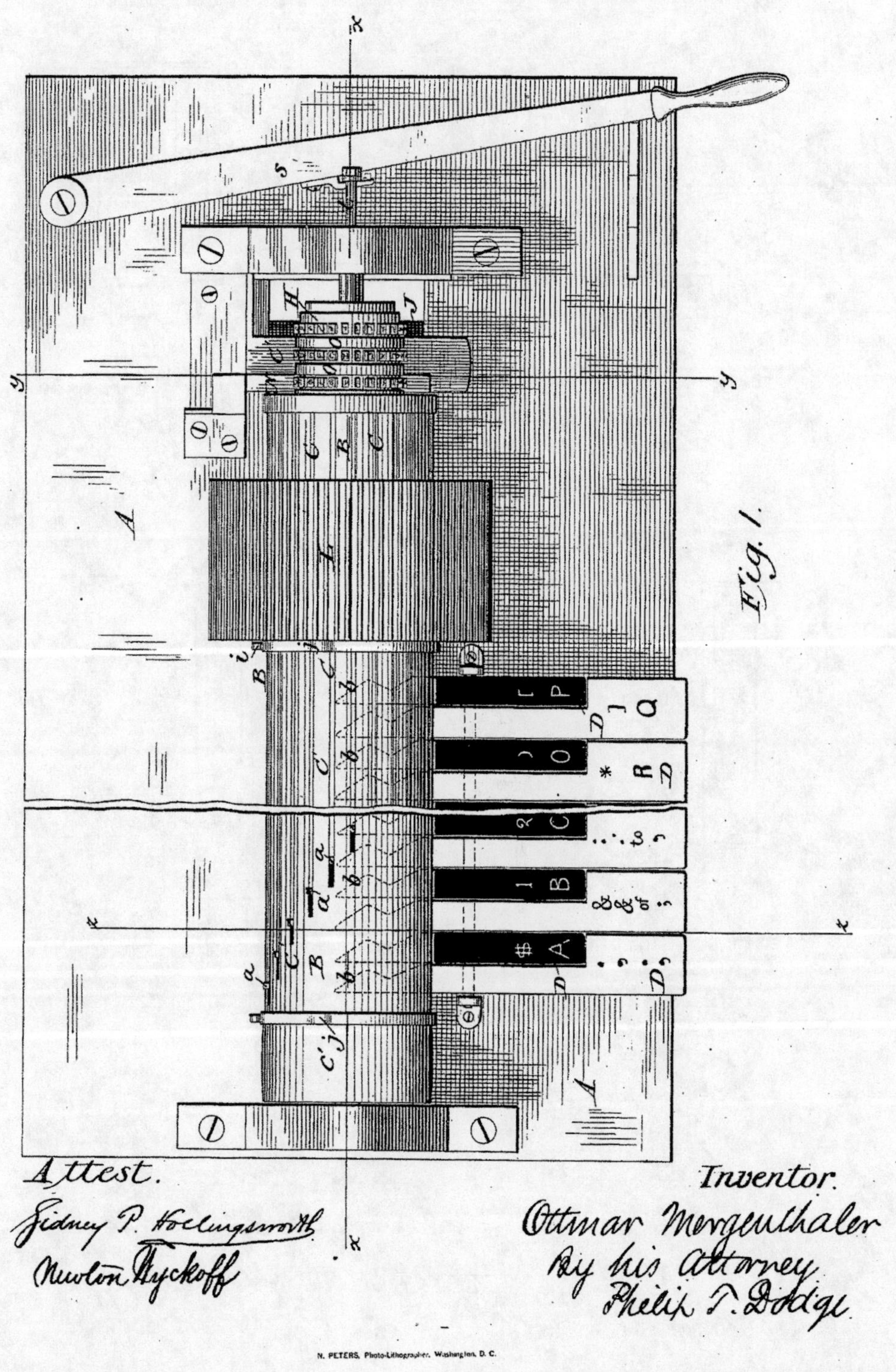

WOWidea! Replace the centuries old tradition of using discreet letters with one line of type! Linotype® was created by Ottmar Mergenthaler to speed up the arduous task of setting thousands of individual letters of the alphabet to form words and sentences for printing. Newspapers embraced this innovation by buying the rights to it.

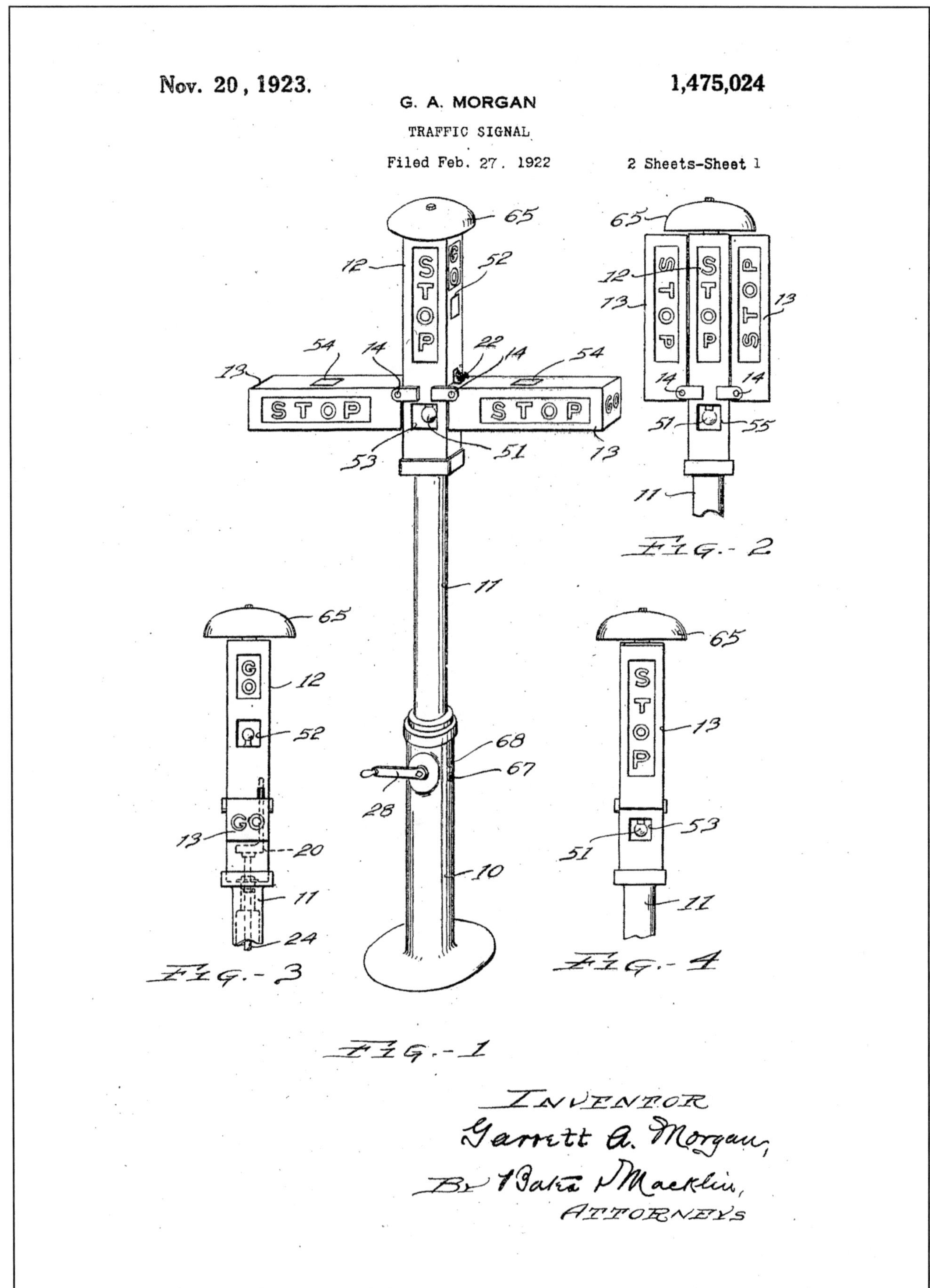

WOWidea! Add "go" lights to stop sign! The son of slaves, Garrett Morgan had already become famous for his invention of a gas mask, saving thousands of lives in World War I. He was inspired to create this more effective traffic control system when he witnessed an awful accident between a motor car and a horse-drawn carriage.

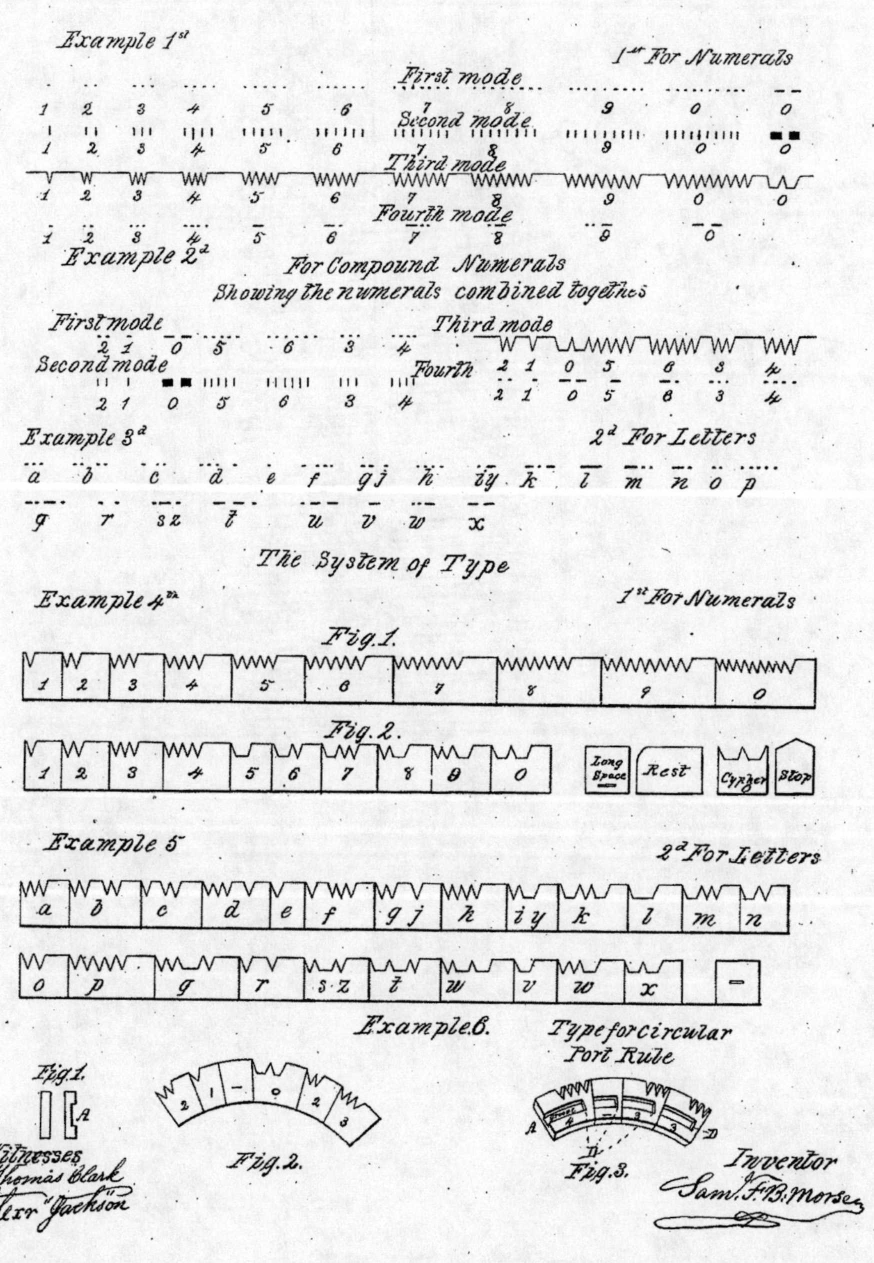

WOWidea! Famous artist paints a picture of success! Samuel F.B. Morse was a portrait painter, but his greatest masterpiece was the ability to transform the hottest technology of the day, electricity, into a workable system to relay information. The telegraph was born along with his ingeniously simple code of dots and dashes.

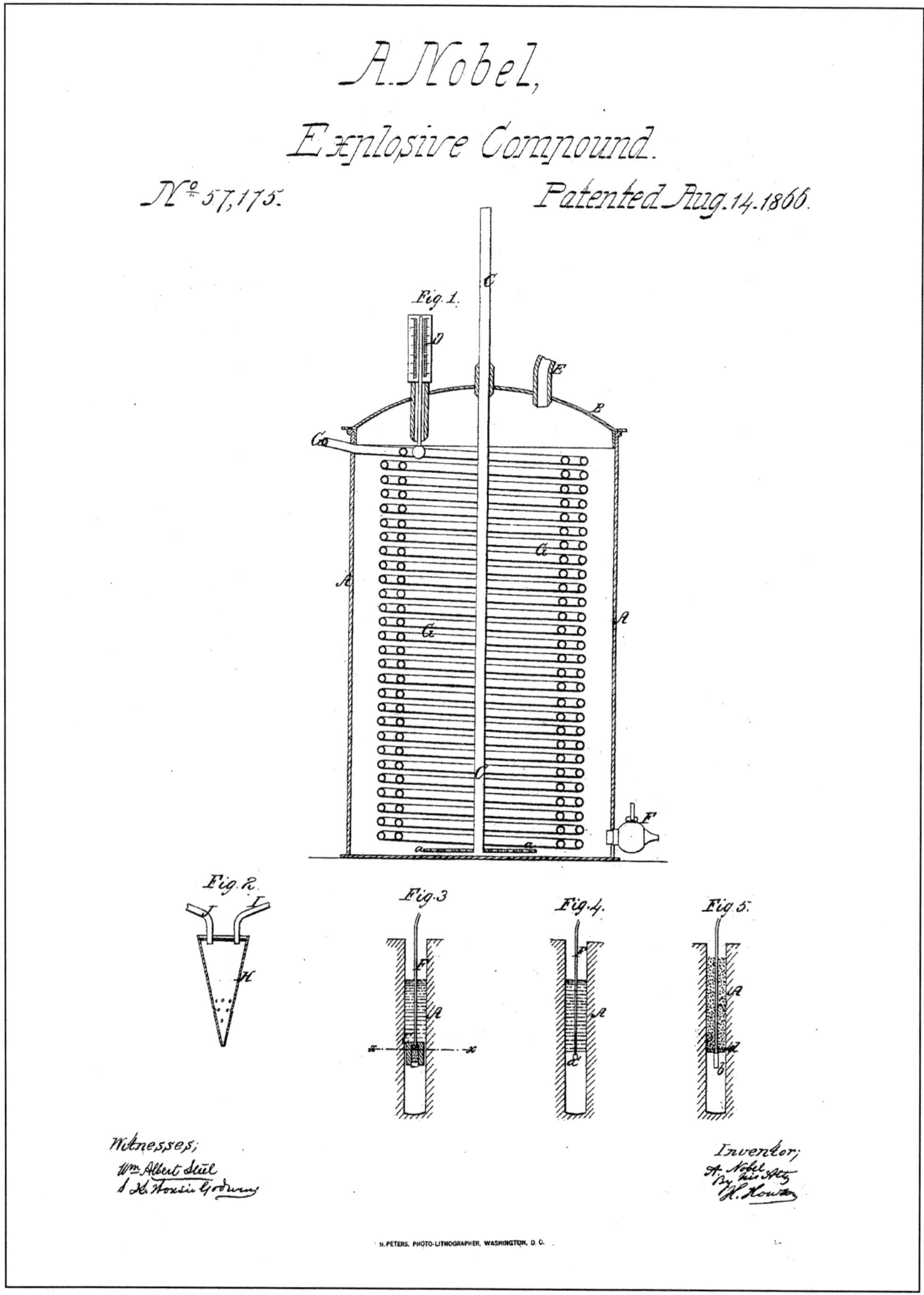

WOWidea! Stabilize it with mother earth! Mixing explosive nitroglycerine with a form of clay, allowed Alfred Nobel to create "Dynamite", something that could be easily handled. Experiments cost him his brother's life, and yet despite the sometimes awful consequences, in the end, his fortune created a positive legacy, the Nobel Prizes.

April 25, 1961 R. N. NOYCE 2,981,877

SEMICONDUCTOR DEVICE-AND-LEAD STRUCTURE

Filed July 30, 1959 3 Sheets-Sheet 1

FIG-1

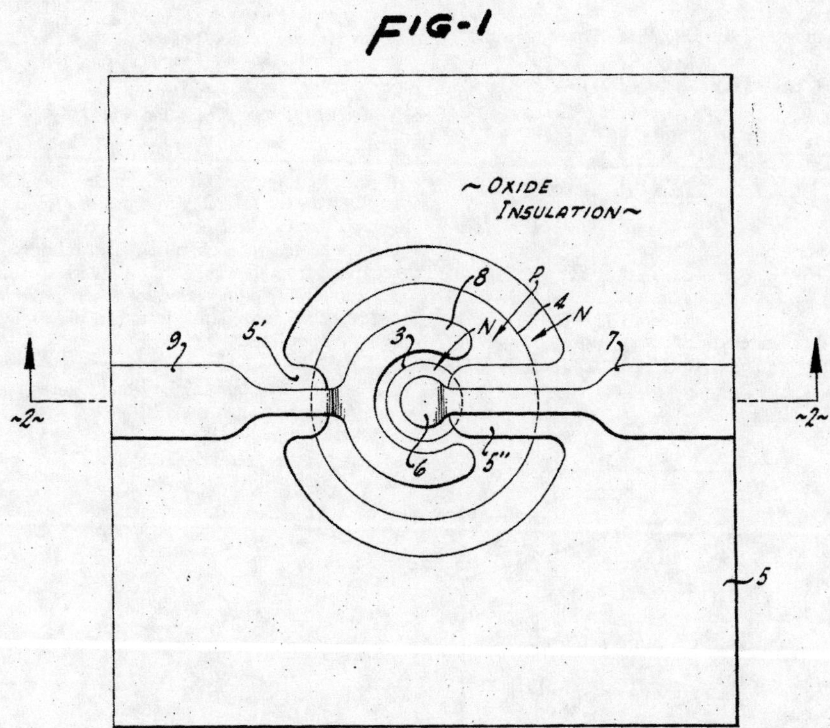

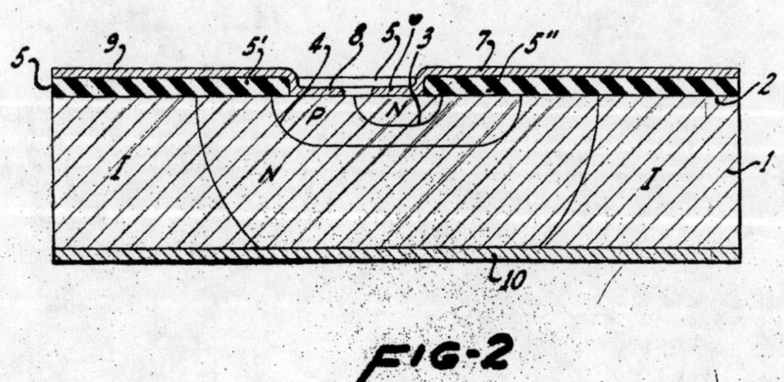

FIG-2

INVENTOR.
ROBERT N. NOYCE
BY Lippincott & Ralls
ATTORNEYS

WOWidea! Cash in the chips! The physical constraints of electronic components was limiting their usefulness as their size outgrew practicality. Leave it to Iowa farm boy, Robert Noyce, to create the first silicon-based crystal chip design that had many transistors etched on it's tiny surface. He and Gordon Moore went on to form Intel®.

United States Patent [19]

Obara

[11] **Patent Number: 4,516,948**

[45] **Date of Patent: May 14, 1985**

[54] **RECONFIGURABLE TOY ASSEMBLY**

[75] Inventor: **Hiroyuki Obara,** Tokyo, Japan

[73] Assignee: **Takara Co., Ltd.,** Tokyo, Japan

[21] Appl. No.: **584,460**

[22] Filed: **Feb. 28, 1984**

[30] **Foreign Application Priority Data**

Sep. 22, 1983 [JP] Japan .... 58-146785[U]

[51] **Int. Cl.**[3] **A63H 17/00**

[52] **U.S. Cl.** **446/95**; 446/97; 446/376; 446/434; 446/487

[58] **Field of Search** 446/95, 94, 93, 97, 446/99, 268, 376, 431, 433, 434, 465, 470, 478, 487

[56] **References Cited**

U.S. PATENT DOCUMENTS

4,206,564	6/1980	Ogawa	446/94
4,382,347	5/1983	Murakami	446/433
4,391,060	7/1983	Nakane	446/94
4,435,916	3/1984	Iwao et al.	446/470

Primary Examiner—Mickey Yu

Attorney, Agent, or Firm—Jackson, Jones & Price

[57] **ABSTRACT**

A reconfigurable toy assembly having foldable portions to allow the toy assembly to simulate a toy combination vehicle having a tractor unit and a trailer unit separatably connected to each other. The tractor unit when separated from the trailer unit is reversibly reconfigurable into a robotic humanoid form, while the trailer unit is reversibly reconfigurable into a play space for the robotic humanoid.

11 Claims, 7 Drawing Figures

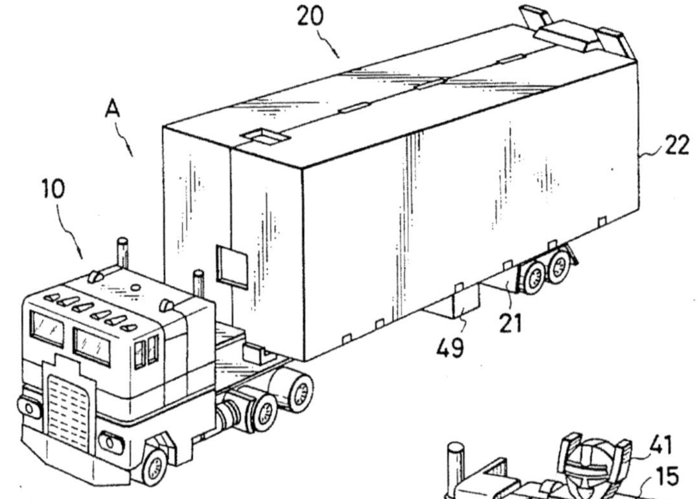

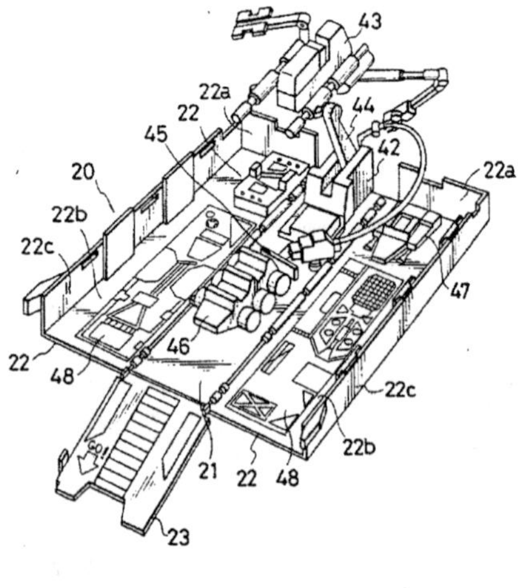

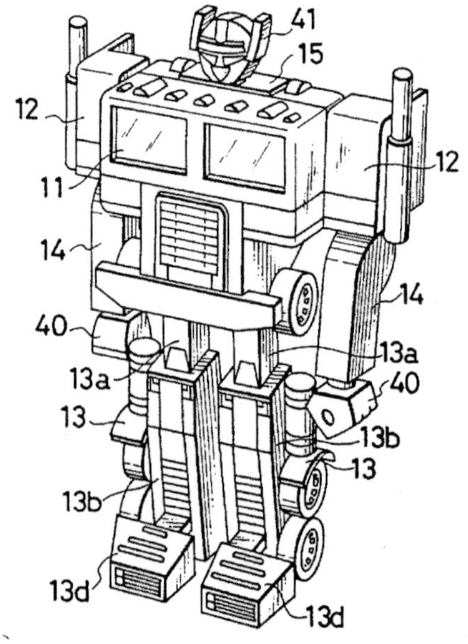

WOWidea! If a truck can become a robot, why can't I dream of becoming anything I want to be? As a must-have toy during the 1980's, Hiroyuki Obara, was one of several inventors busily at work for Takara, the Japanese subsidiary of Hasbro. Saturday morning cartoons and, eventually a movie, rounded out the Transformers® craze.

March 1, 1938. F. J. OSIUS **2,109,501**

DISINTEGRATING MIXER FOR PRODUCING FLUENT SUBSTANCES

Filed March 13, 1937 2 Sheets-Sheet 1

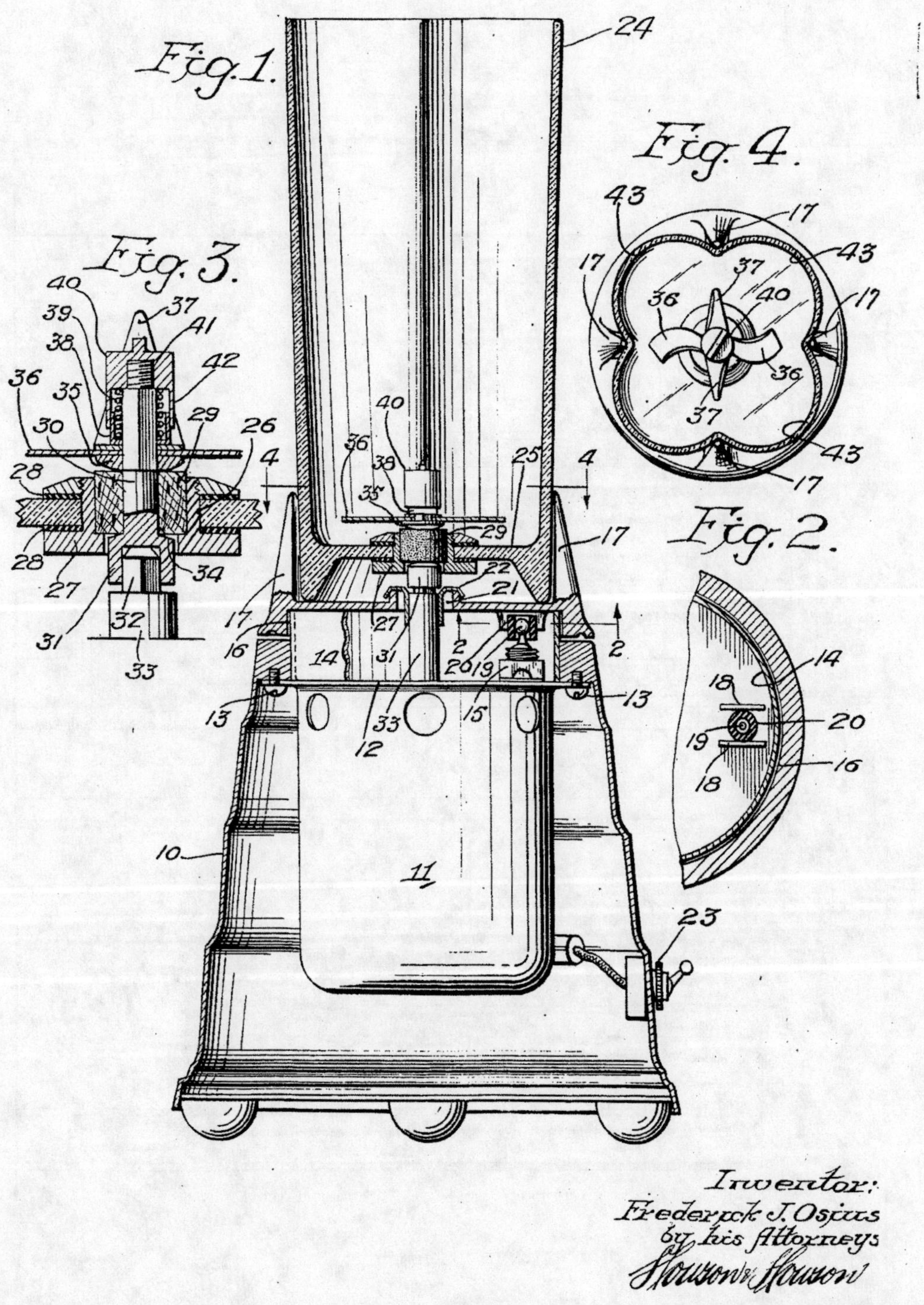

WOWidea! A famous name never hurts! Fred Osius wouldn't have been as successful had he not gone to the dressing room of bandleader Fred Waring and sold him on the idea of blended banana shakes. The Blendor® became a kitchen staple, set the stage for food processors and even helped Jonas Salk prepare his polio vaccine.

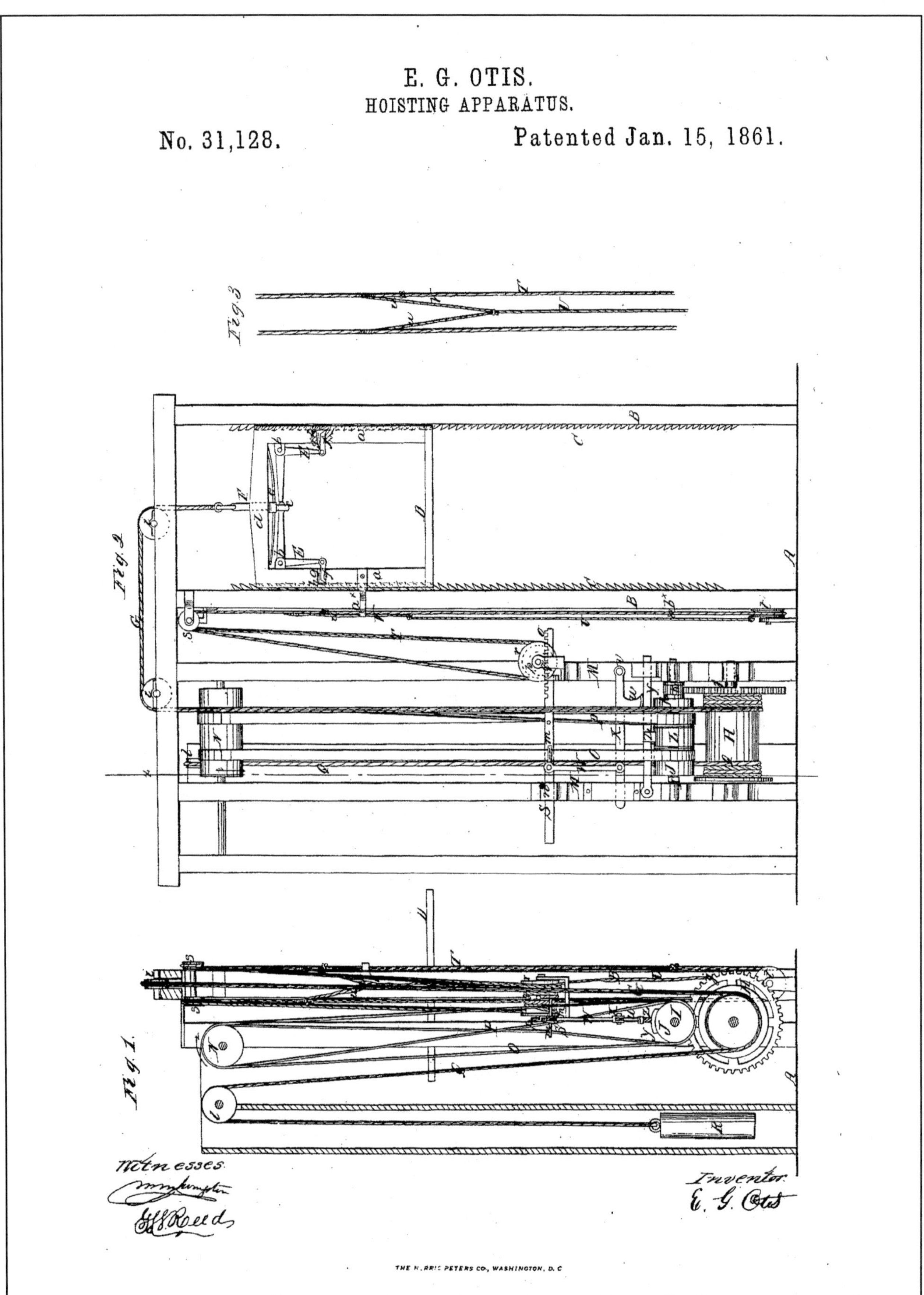

WOWidea! Safety-catch caught success! Elisha Graves Otis, a master mechanic who suffered through several illnesses in his life, envisioned this invention while working on the site of his employer's new building. Passing away in the year this patent was issued, his sons carried the torch by establishing the Otis Elevator Company.

4 Sheets—Sheet 1.

N. A. OTTO.

GAS-MOTOR ENGINES.

No. 194,047. Patented Aug. 14, 1877.

Fig. 1.

Fig. 4.

Witnesses

B. C. Pole

N. B. Whitman

Inventor

N. A. Otto by

C. S. Whitman

atty

WOWidea! Thinks he can sell a few! Nikolaus Otto, a salesman from Germany, patents this economical four-cycle engine in the U.S., breaking a patent monopoly with the help of Ford Motor Co., yet loses an important European patent battle himself. A former Otto employee? Gottlieb Daimler, later names his car for his daughter, Mercedes.

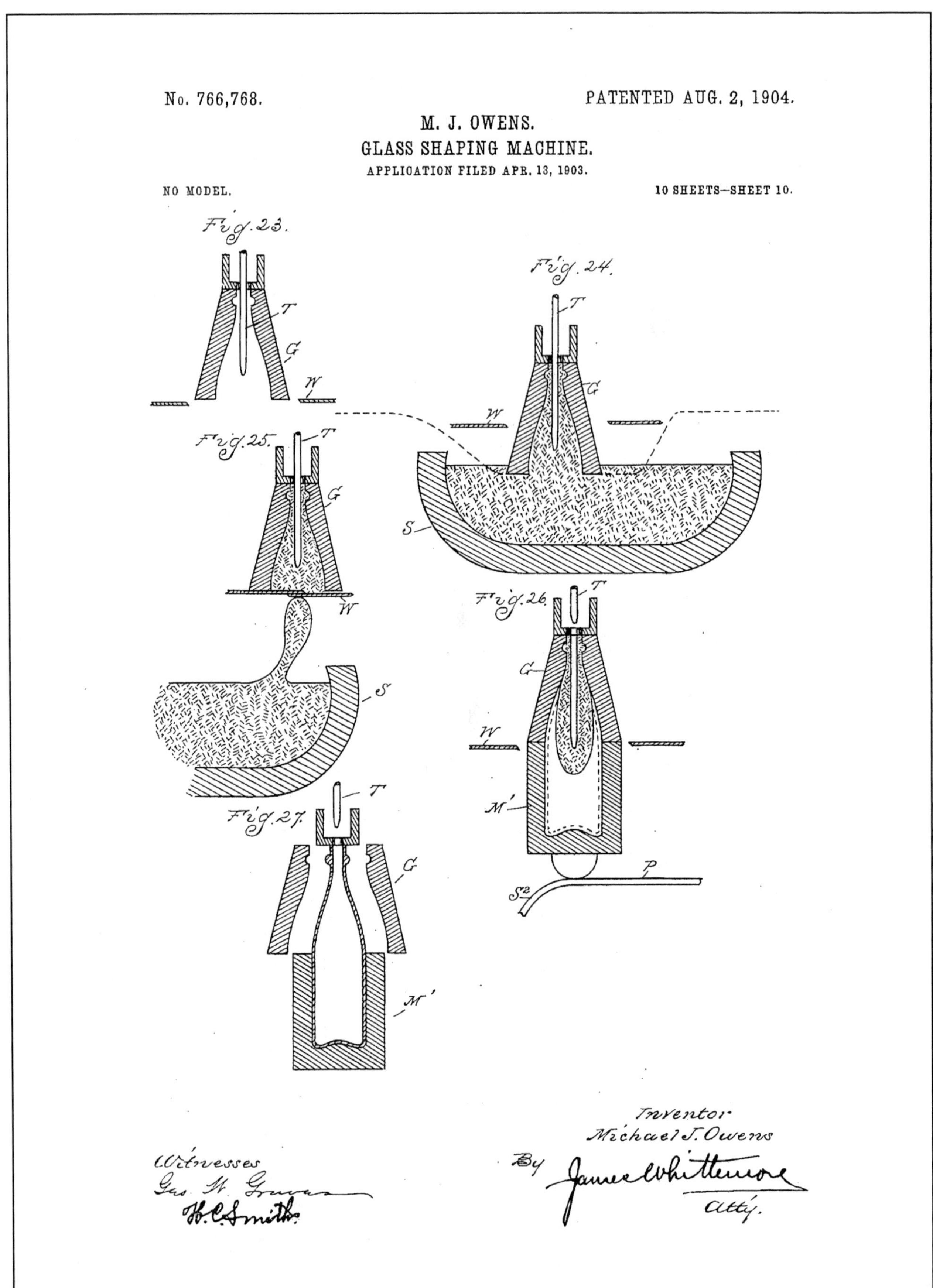

WOWidea! A vacuum, molten glass, and compressed air help create 2,500 bottles an hour! For over two thousand years, hand-blown glass containers, painstakingly made, were all we had. Then came Michael Owens, the son of a coal miner, with his automated bottle factory. It's among 45 patents he was awarded in the glass industry.

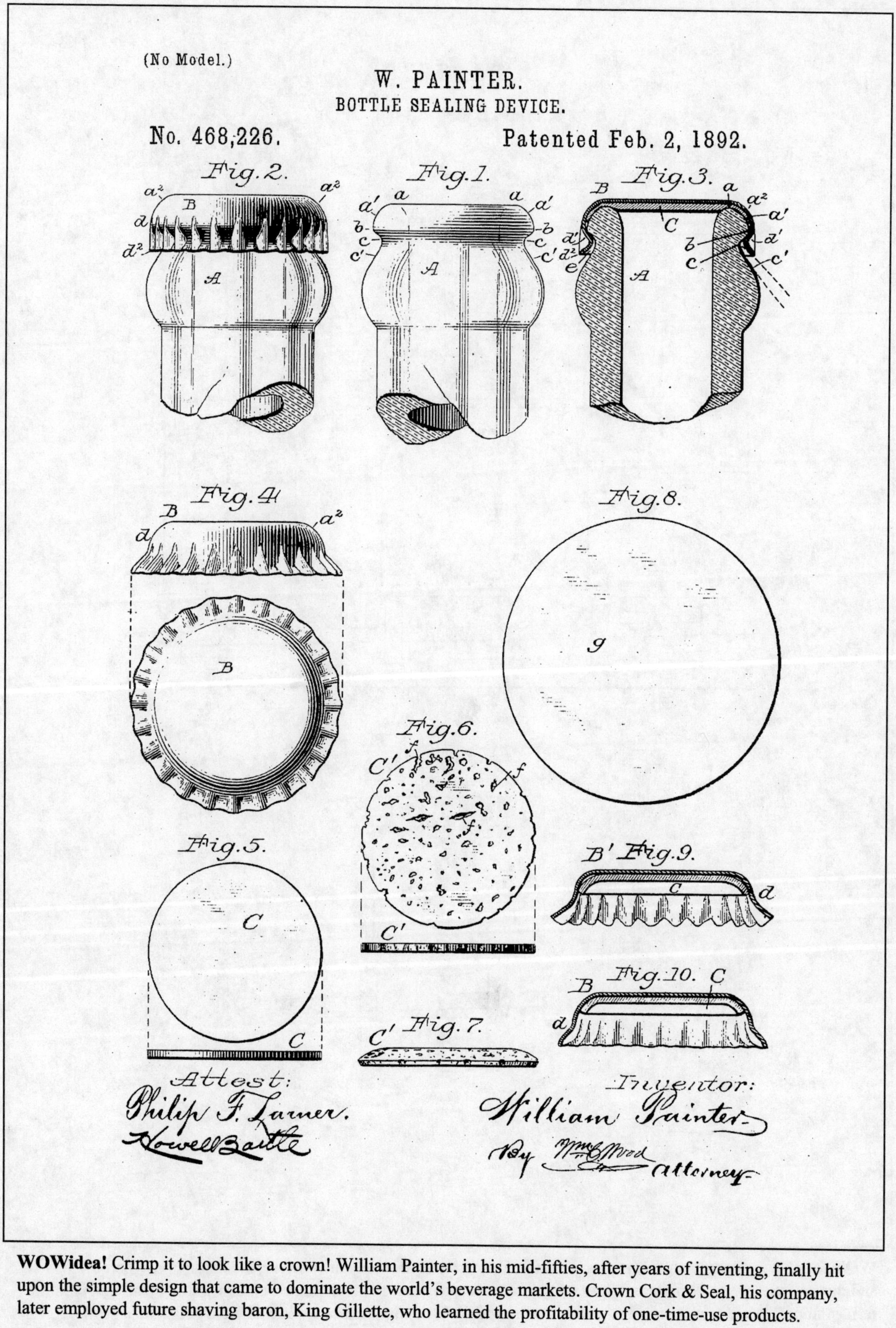

(No Model.)

W. PAINTER.

BOTTLE SEALING DEVICE.

No. 468,226. Patented Feb. 2, 1892.

Fig. 2. Fig. 1. Fig. 3.

Fig. 4. Fig. 8.

Fig. 6.

Fig. 5. Fig. 9.

Fig. 10.

Fig. 7.

Attest: Philip F. Larner. Howell Bartle

Inventor: William Painter By Wm. C. Wood Attorney

WOWidea! Crimp it to look like a crown! William Painter, in his mid-fifties, after years of inventing, finally hit upon the simple design that came to dominate the world's beverage markets. Crown Cork & Seal, his company, later employed future shaving baron, King Gillette, who learned the profitability of one-time-use products.

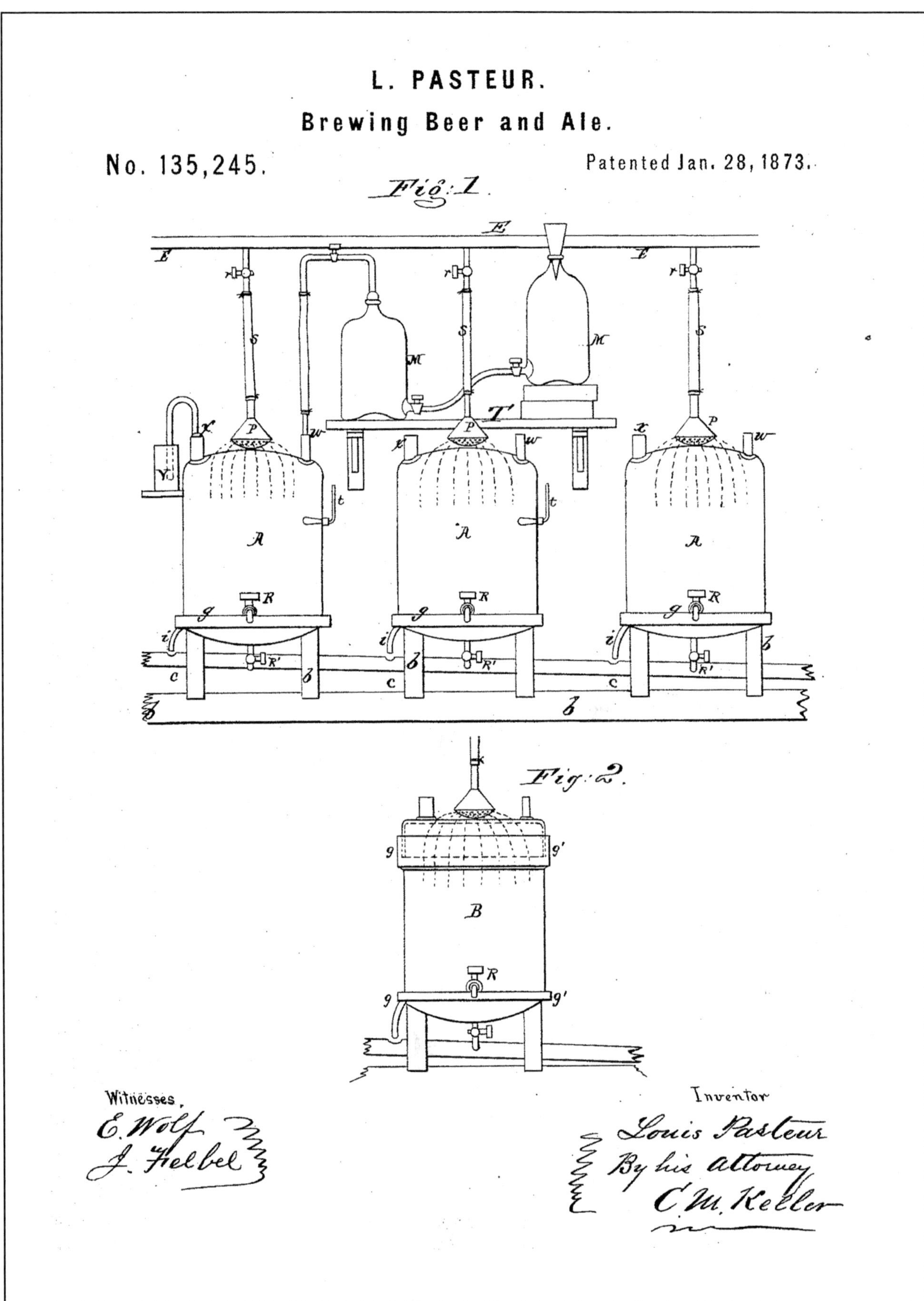

WOWidea! Living organisms, if properly managed, add to the taste! Louis Pasteur, the "father" of microbiology and the inventor of "pasteurization", the heating of milk and other foods to preserve them, was also an innovator in fermentation. This patent depicts his careful control of the brewing process when making a delicious pale ale.

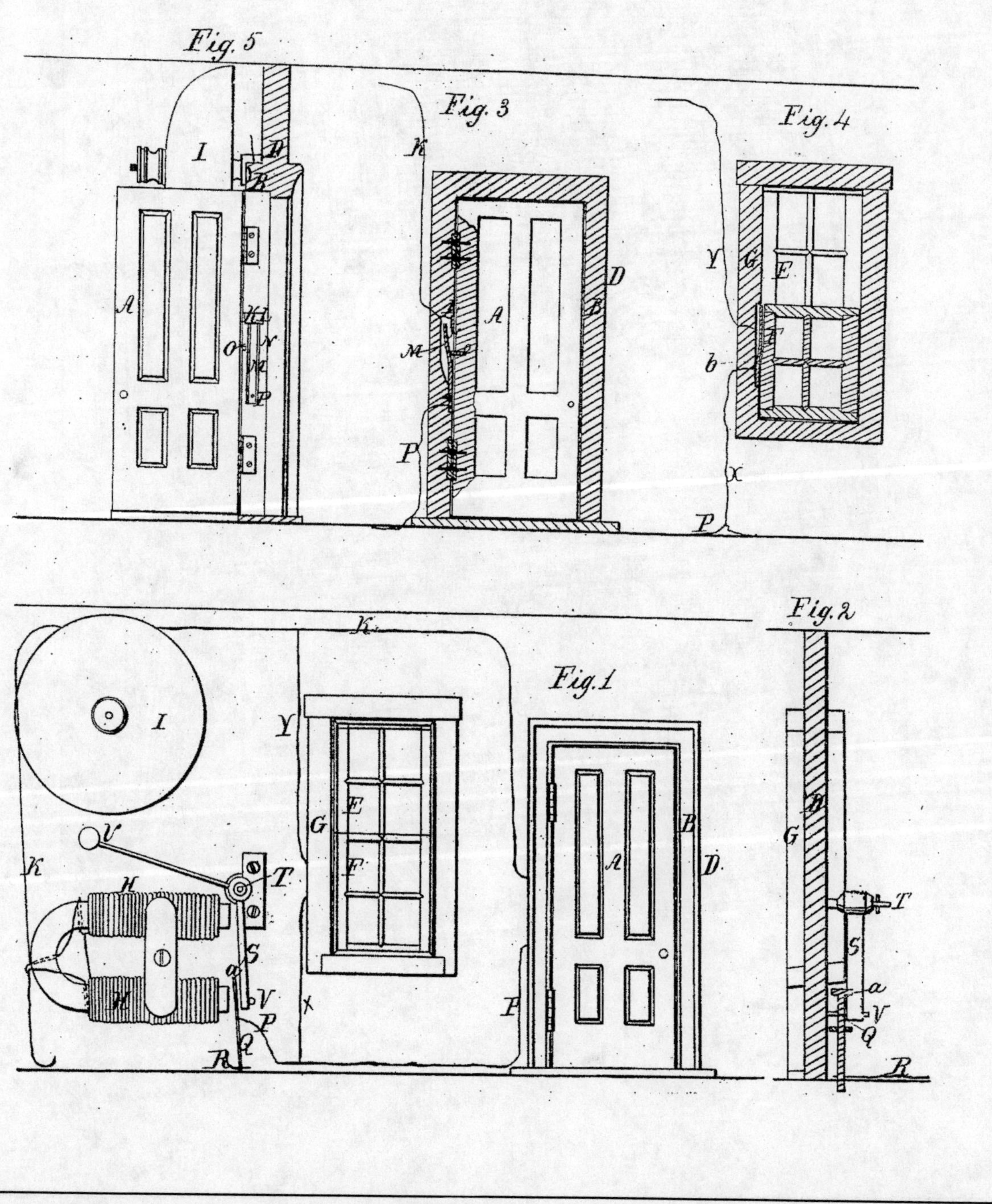

WOWidea! Bad guys and bells don't mix! Augustus Pope, using the latest technology of the day, electricity, was out to improve upon the ancient necessity of protecting your home from unwanted intruders. Even Thomas Edison, as a young entrepreneur, and long before his first light bulb was lit, advertised that he installed burglar alarms.

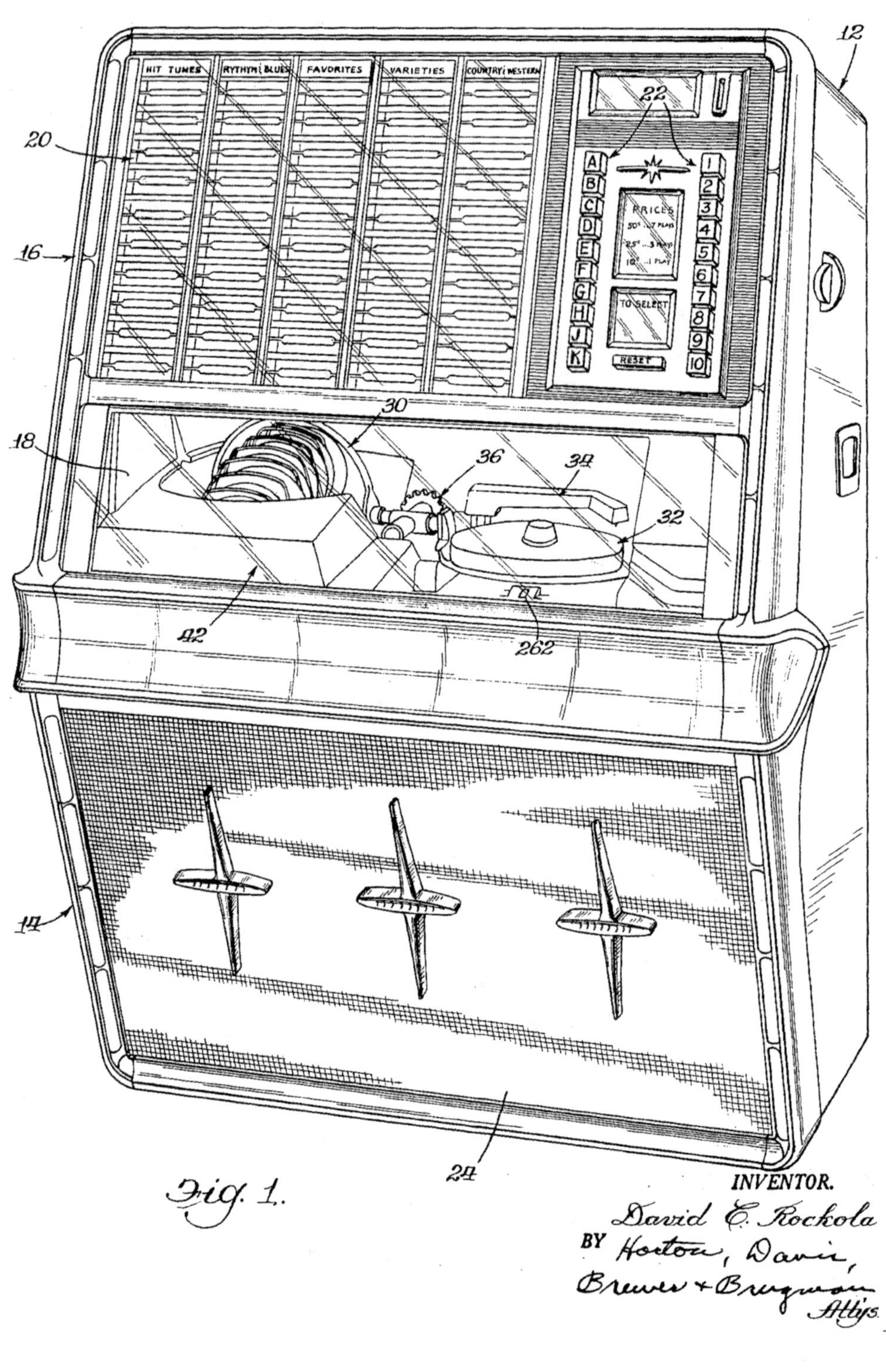

WOWidea! If the name fits, build it! At a relatively young age, David Rockola started building coin-operated games, and by the 1930s had decided to join the competitive jukebox marketplace. When Rock n' Roll came on the music scene in the 1950s, Rock-Ola Manufacturing was ready, and came to dominate the 45rpm record era.

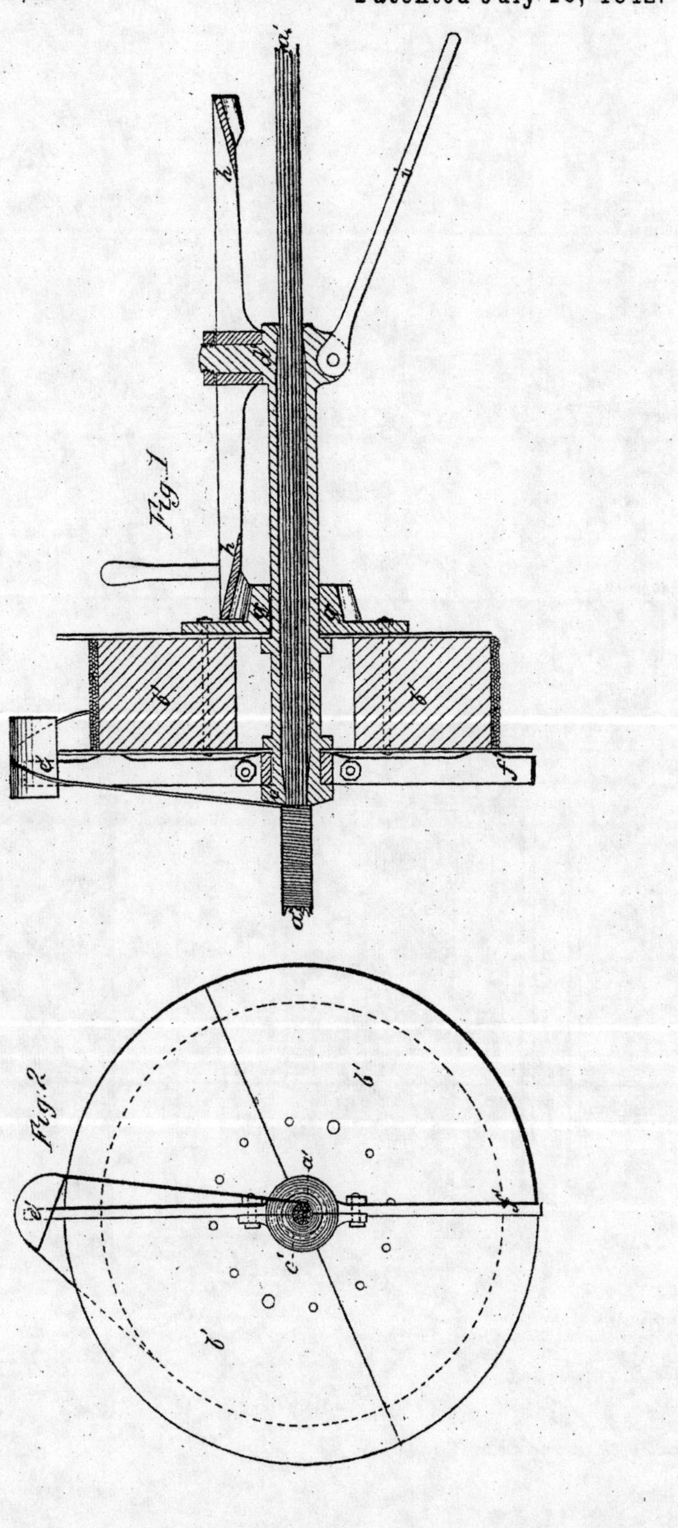

WOWidea! Wire rope can give so much! John Roebling pioneered the use of twisted steel wire rope, first on a Pennsylvania canal, then for telegraph cables and finally, with his family, built the Brooklyn Bridge, a tribute to great American engineering. Even the first Otis elevator used Trenton, NJ-made Roebling suspension cables.

WOWidea! Solid upbringing resulted in solid construction! Lillian Russell, a mid-west farm girl from Clinton, Iowa, became the beautiful soprano opera singer and actress of the late 19th and early 20th centuries. Her travel needs led to this combination dresser-trunk and dressing room, built solidly enough to remain intact while touring.

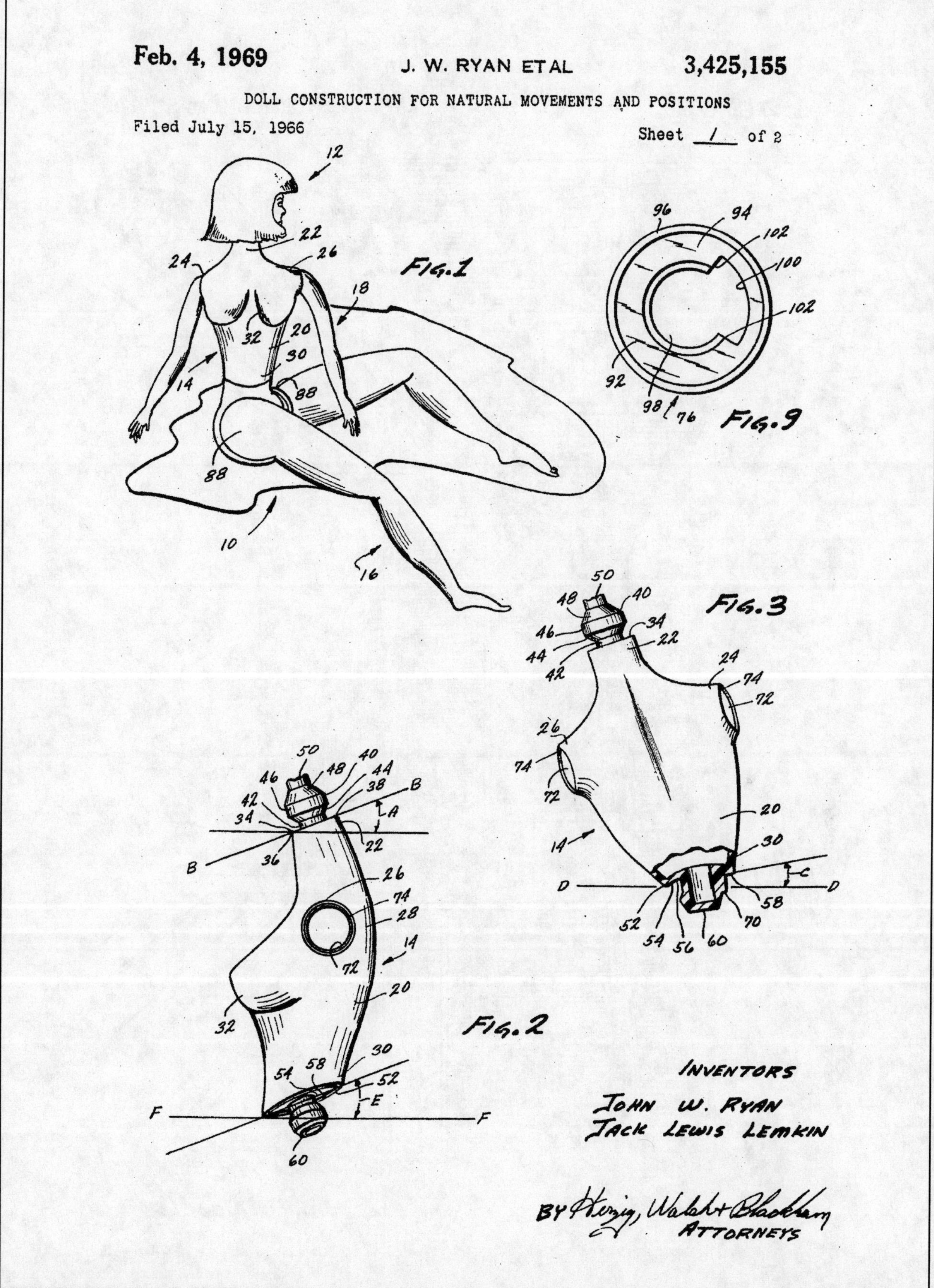

WOWidea! Doll with breasts deemed important to a little girl's self-esteem! Introduced at a toy fair in New York City in 1959 by Ruth Handler, Barbie®, named after her daughter, became a most sought-after doll. John Ryan and Jack Lemkin gave her an updated design with movable body parts which added to the realism and playtime fun.

WOWidea! Let the customer help themselves! Until Clarence Saunders came on the scene, most stores had clerks which brought the goods out to you. The idea of a self-service "maze" centered on higher sales and lower costs, yet customers loved the convenience. His Piggly Wiggly® markets helped create a nation of shopaholics.

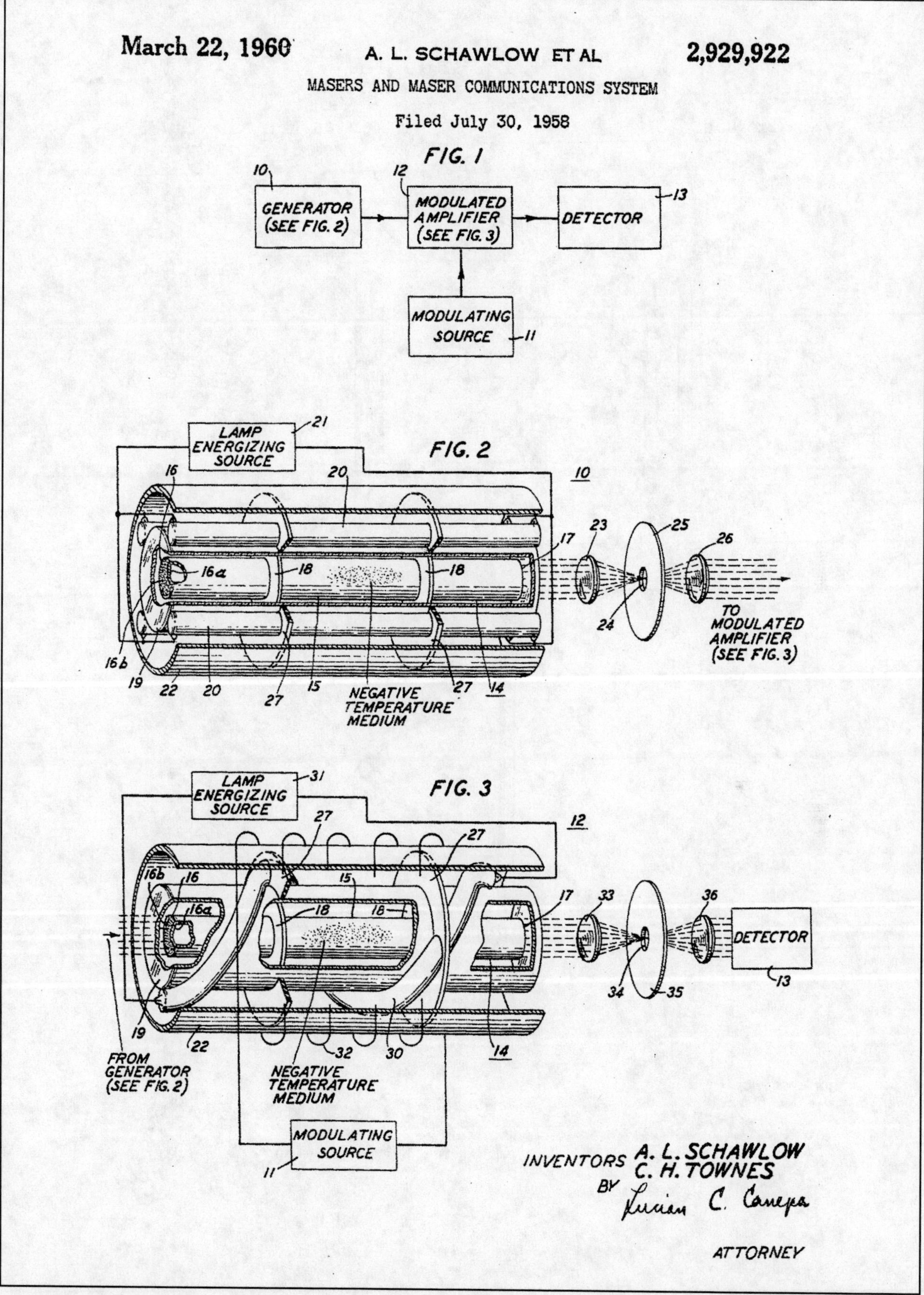

WOWidea! Eureka moment came while resting on a park bench! Among the many people who contributed to the theoretical and practical realization of Microwave Amplification by Stimulation Emission of Radiation (MASER) and its more well known light version, the LASER, were brothers-in-law Arthur Schawlow and Charles Townes.

(No Model.)

M. P. SCHOOLEY.

PAPER CLIP OR HOLDER.

No. 601,384. Patented Mar. 29, 1898.

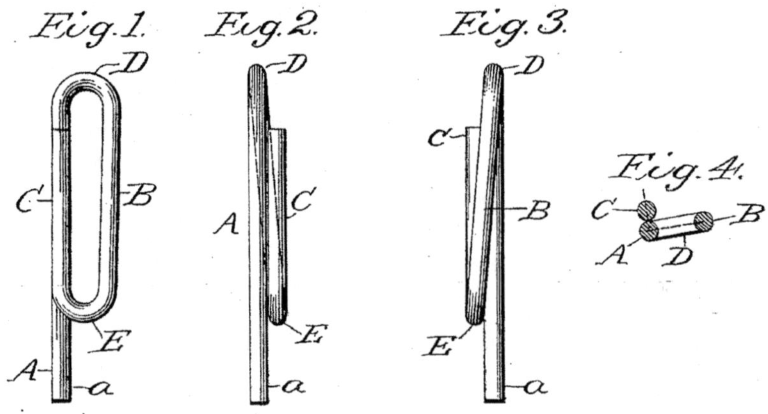

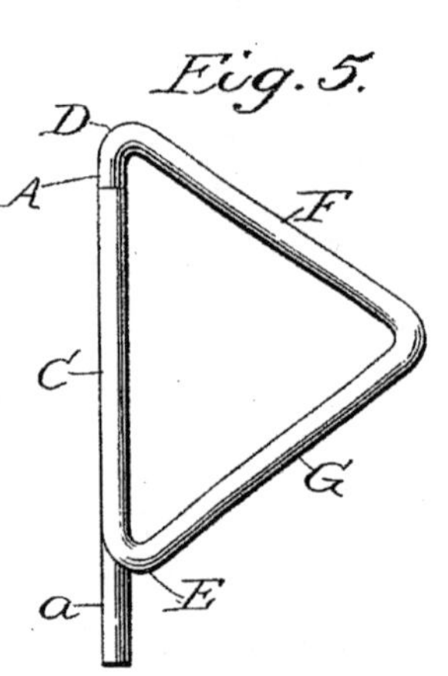

Attest;
C. C. Burdine.
D. E. Burdine.

Inventor;
Matthew P. Schooley,
by Dodge and Sons,
Att'ys.

THE NORRIS PETERS CO., PHOTO-LITHO., WASHINGTON, D. C.

WOWidea! It would bind more than papers! Matthew Schooley is credited with the first American patent for the paperclip, one similar in shape and size to today's. Interestingly, Johann Vaaker, of Norway, created one similar in 1899 and years later, his countrymen pinned paperclips to their lapels to show unity against Nazi occupation.

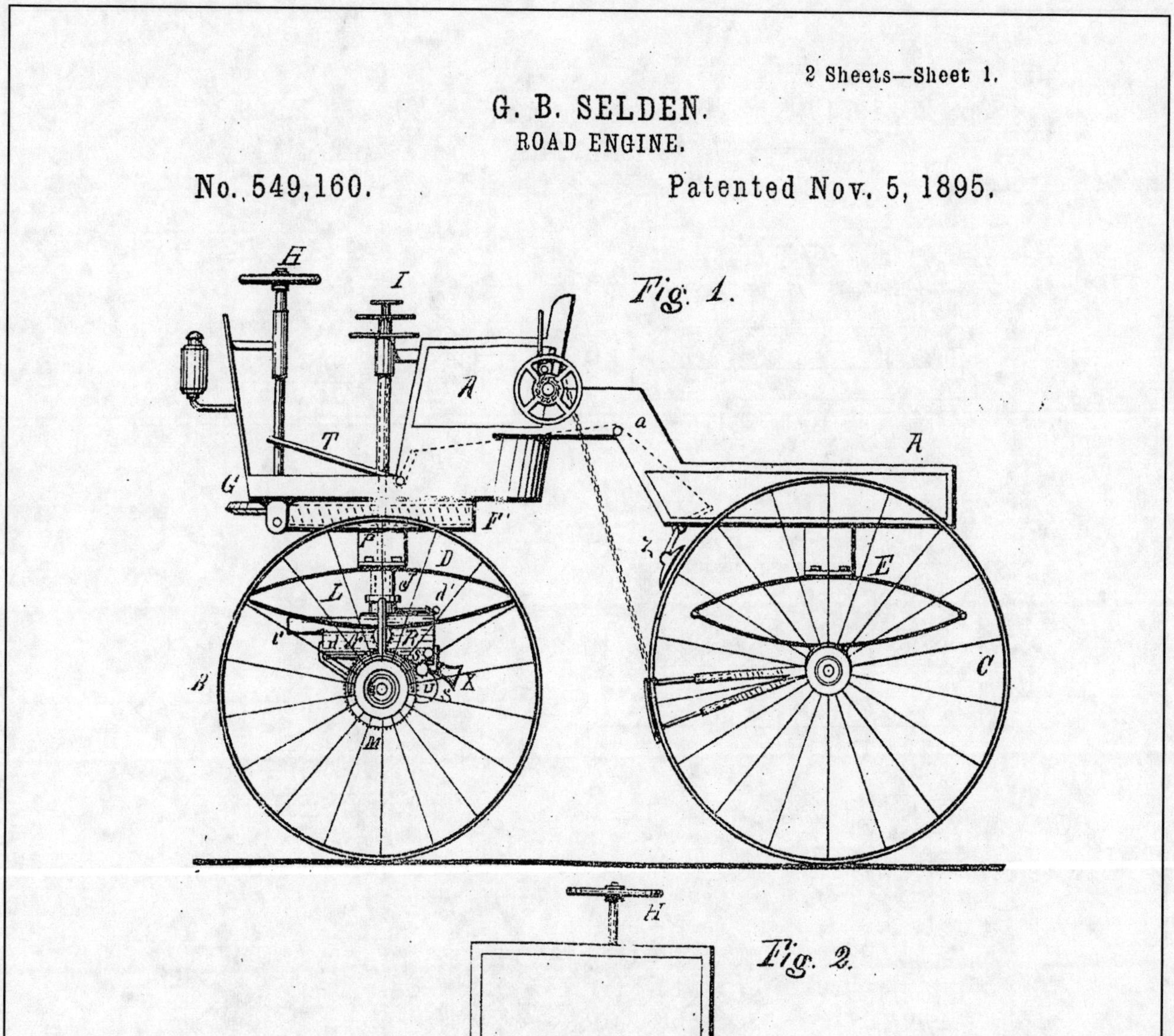

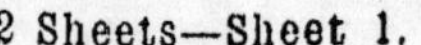

2 Sheets—Sheet 1.

G. B. SELDEN.

ROAD ENGINE.

No. 549,160. Patented Nov. 5, 1895.

Fig. 1.

Fig. 2.

WITNESSES

W. Mc Kelway, Jr.

Geo. Eastman

INVENTOR

Geo. B. Selden

WOWidea! Why manufacture it, license it for big profits! George Selden, a patent attorney, was first to combine a two-cycle internal combustion engine with a carriage and even had George Eastman witness his patent application. Then, with bureaucratic delays, he collected substantial royalties from others until Henry Ford broke the monopoly.

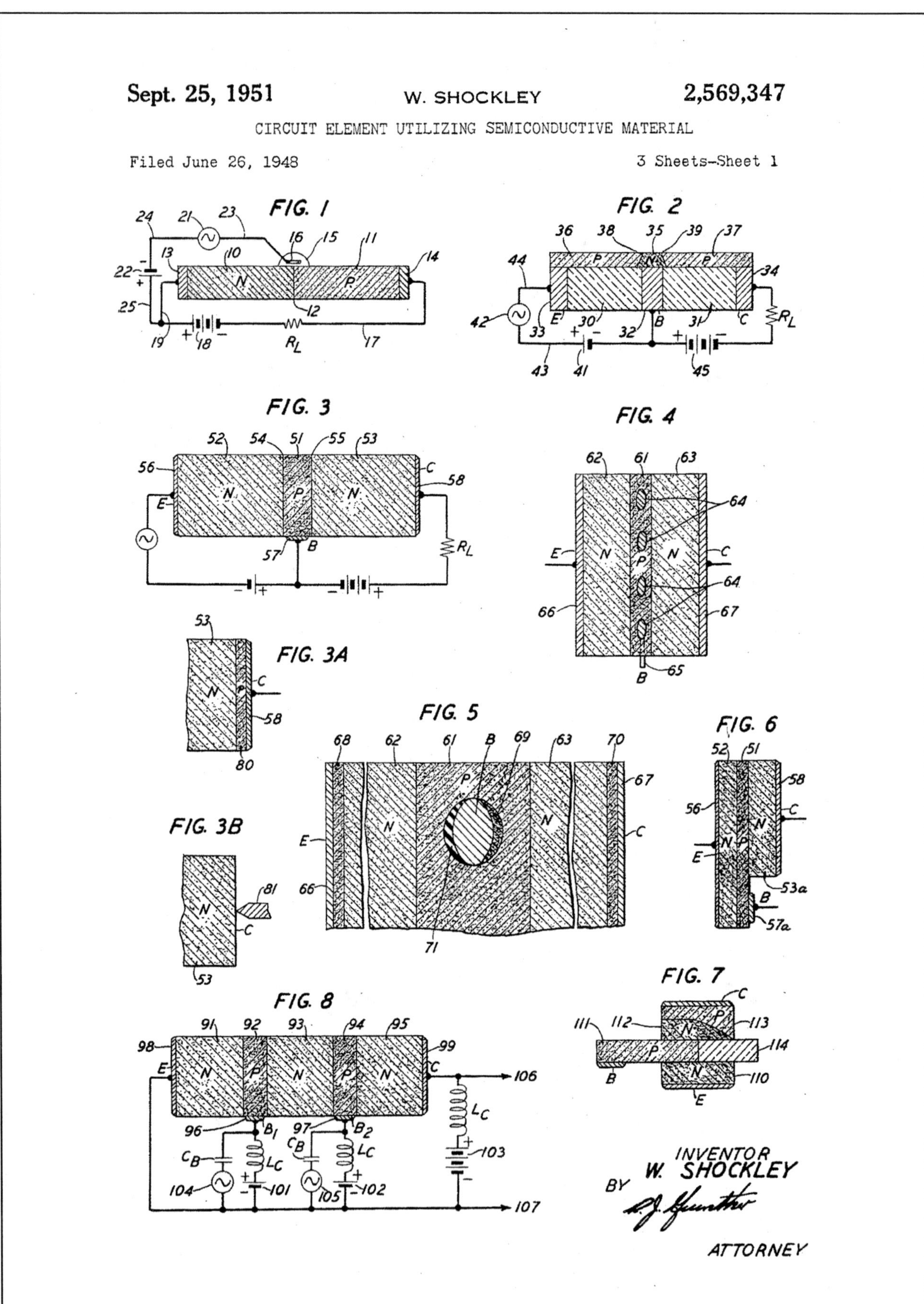

WOWidea! Silicon, part insulator, part conductor, let's call it a semiconductor! The all-important transistor was perfected through the Nobel prize-winning research of William Shockley, with earlier help from Messrs Brattain and Bardeen. Shockley then went to California to start a company, and unwittingly, the "Silicon Valley".

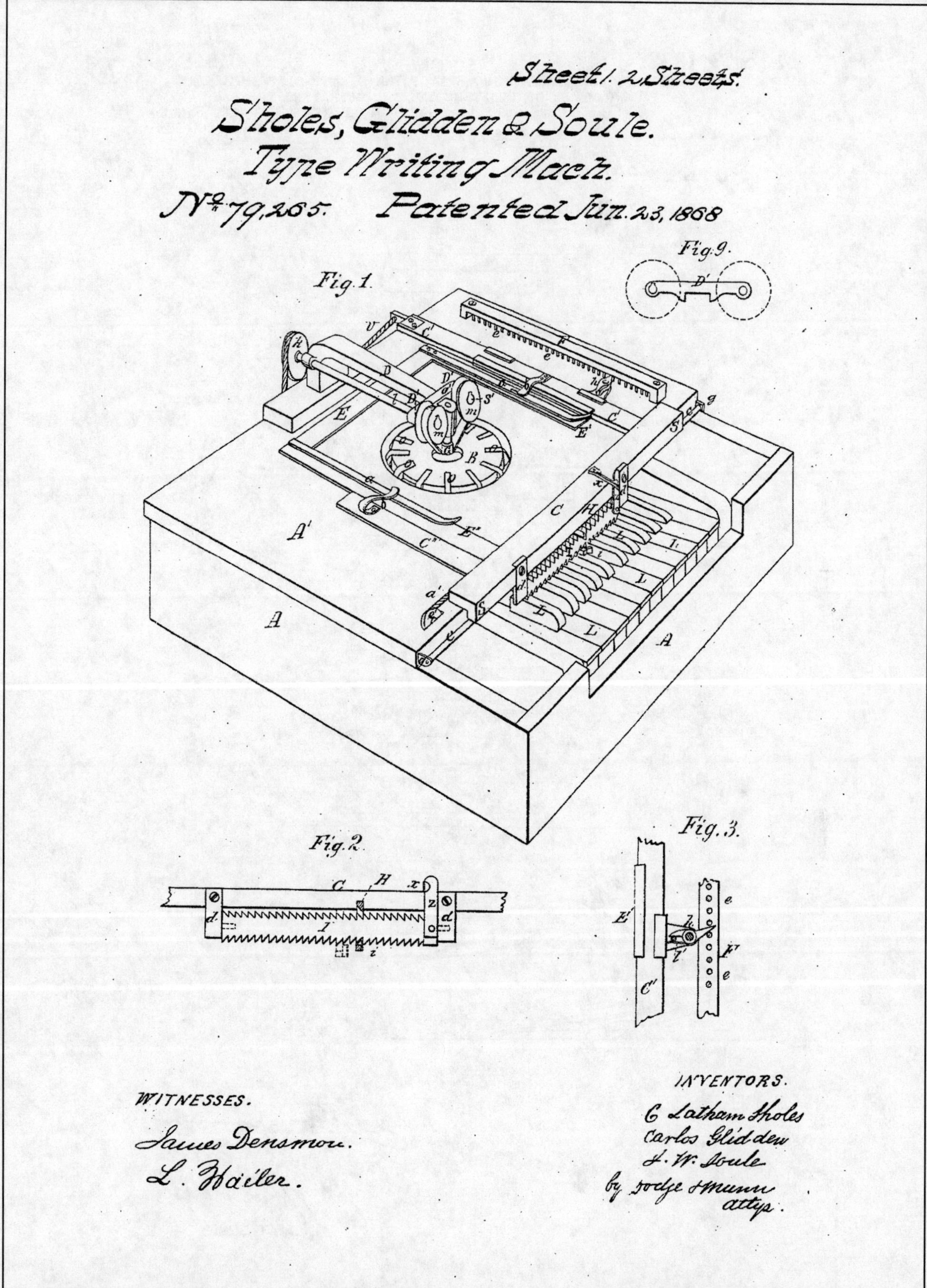

WOWidea! He could have chosen royalties! Christopher Sholes had been a Wisconsin State senator and was even appointed by Lincoln to a federal post. Along with his co-inventors, he created the first practical typewriter. Yet, selling his rights to Remington Arms Co., for one lump sum of cash, he never profited again from its later success.

March 8, 1932. I. SIKORSKY 1,848,389

AIRCRAFT, ESPECIALLY AIRCRAFT OF THE DIRECT LIFT AMPHIBIAN TYPE AND MEANS OF CONSTRUCTING AND OPERATING THE SAME

Original Filed Feb. 14, 1929 8 Sheets-Sheet 1

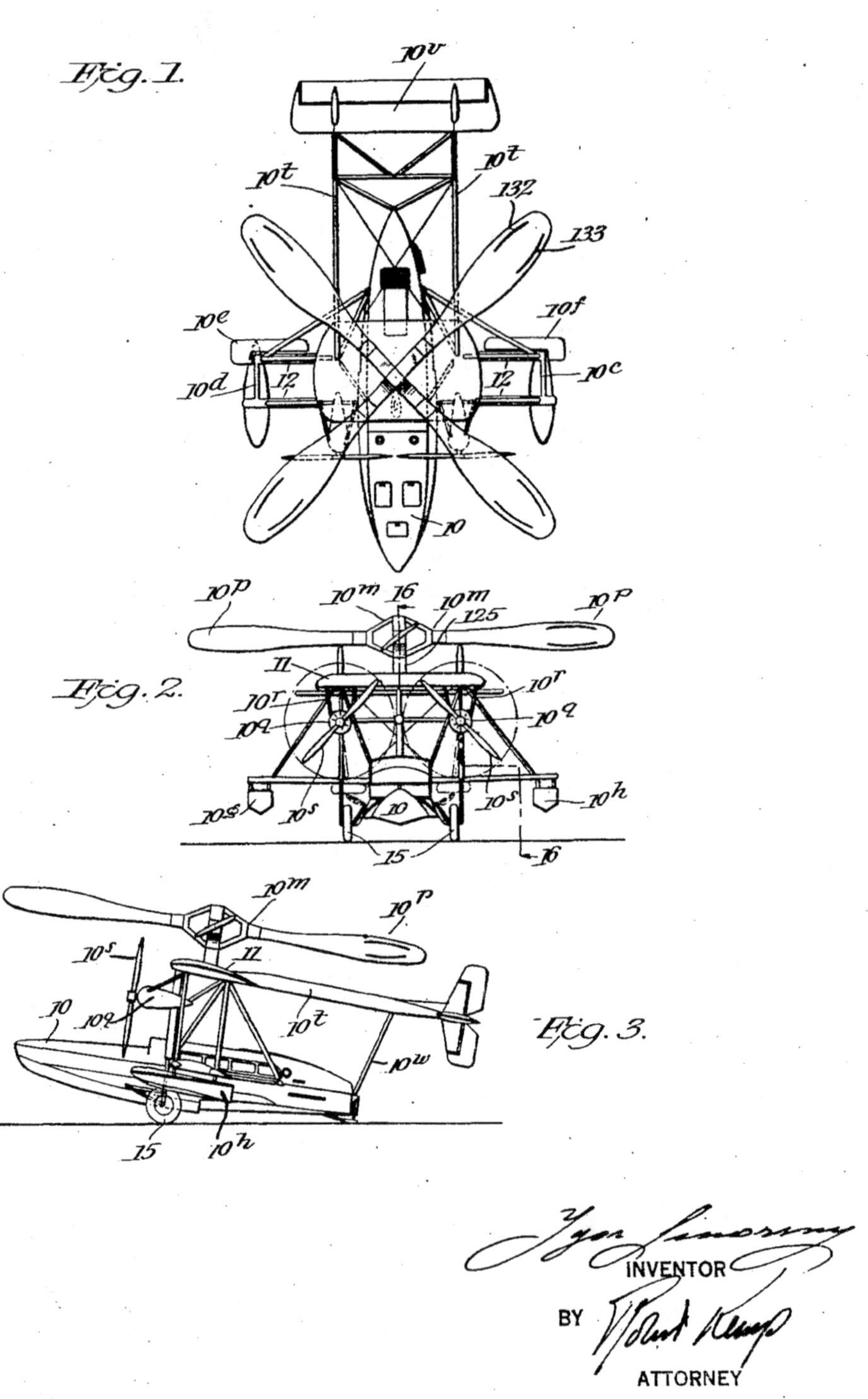

WOWidea! Brief encounter creates life-long endeavor! Ivan Sikorsky, as a young boy growing up in Russia learns of Leonardo da Vinci's design for a helicopter and thus begins his quest. After years of refinement, Sikorsky becomes the world's leading supplier of helicopters thanks to financial backing from virtuoso Serge Rachmaninoff.

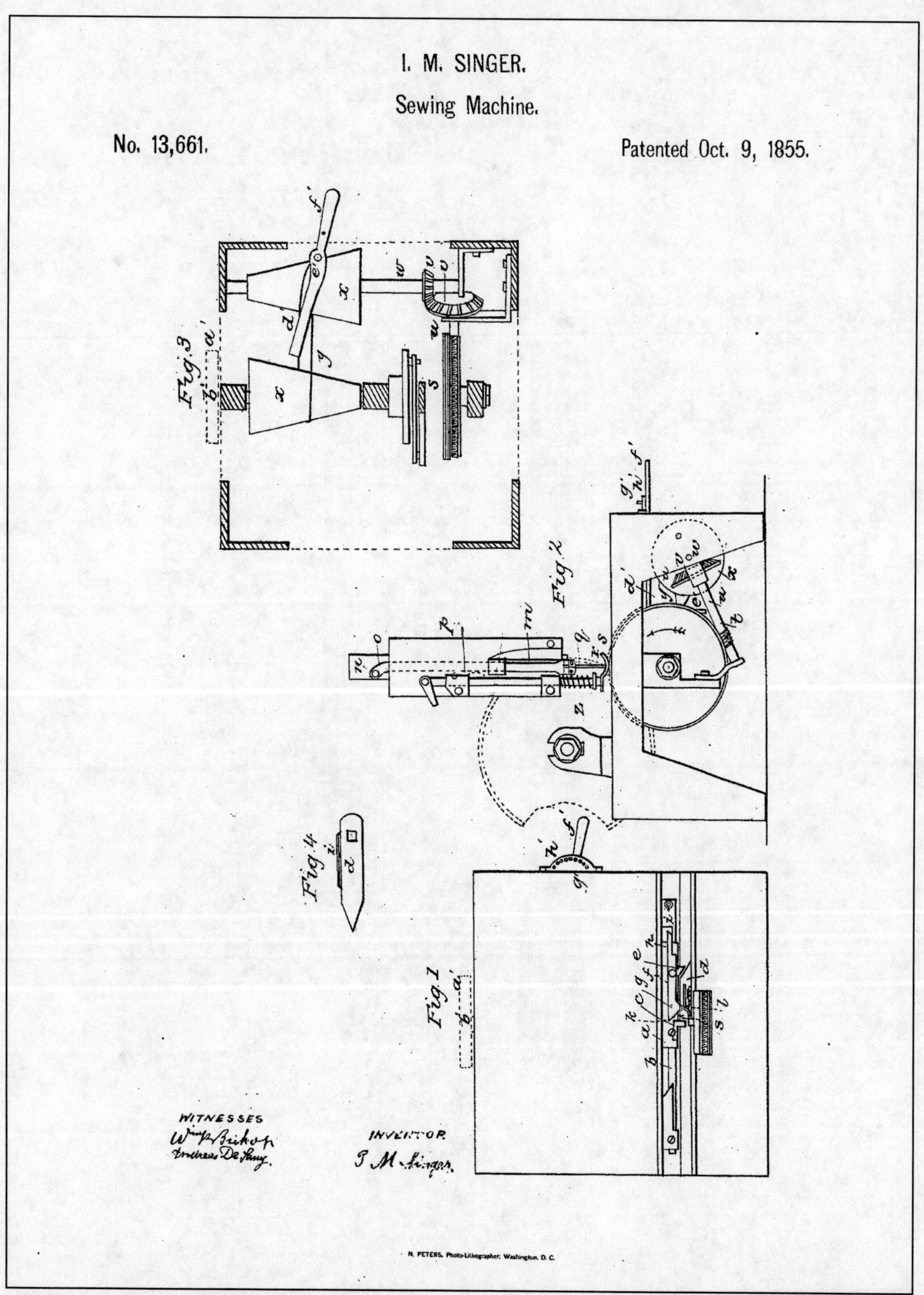

WOWidea! Thread from the shuttle below and needle above, is what made this stitch so lasting! Although Elias Howe is the acknowledged inventor of the first practical sewing machine in 1843, it was Isaac Singer who created one of the first and greatest collaborations of manufacturers specifically designed to monopolize a marketplace.

2 Sheets—Sheet 1.

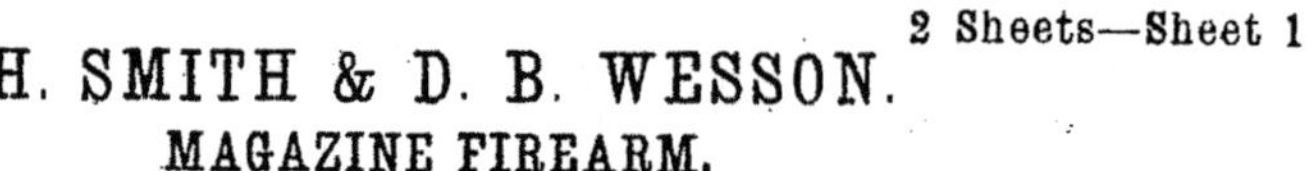

H. SMITH & D. B. WESSON.

MAGAZINE FIREARM.

No. 10,535. Patented Feb. 14, 1854.

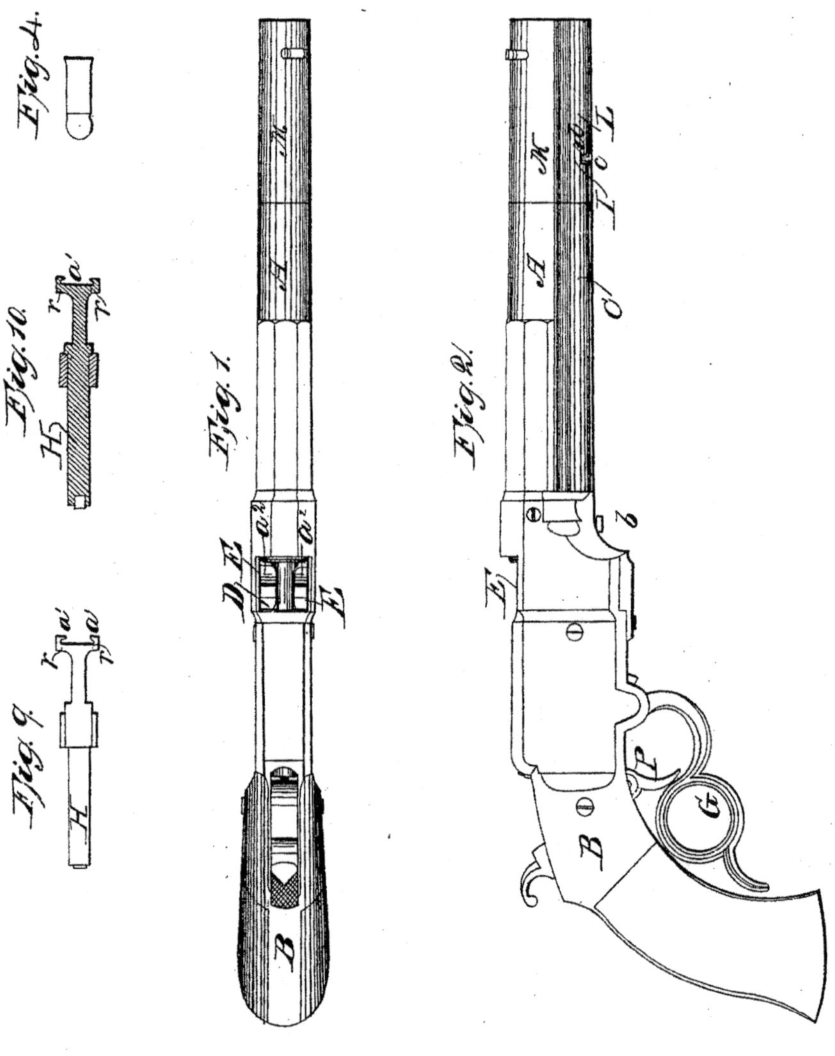

THE NORRIS PETERS CO., PHOTO-LITHO., WASHINGTON, D. C.

WOWidea! Something in their genes! Horace Smith and Daniel Wesson were both sons of machinists and, as partners, had their first success, this famous "Model No. 1" Smith & Wesson firearm. This breech loader chambered for a revolutionary new cartridge, with only minor changes, still exists today as the familiar .22 rim-fire.

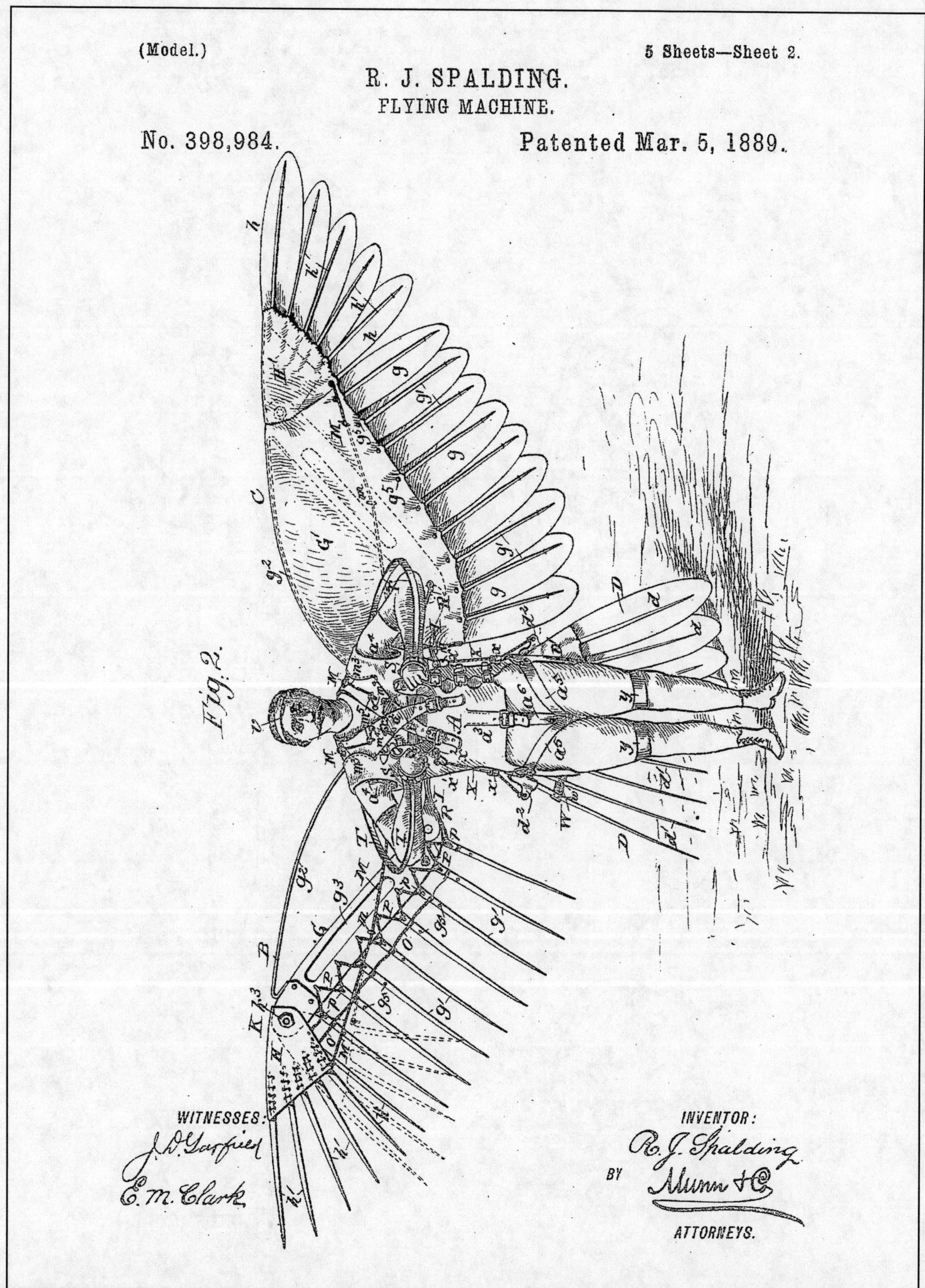

WOWidea! Trial and error are the roots of success! The dreams of Reuben Spaulding and his winged man are long lost to time, but most would agree that without the efforts of would-be inventors like him, the progress of our world would never have gotten off the ground. Who knows what might happen if you spread your creative wings?

Oct. 11, 1966 S. F. SPEERS ET AL 3,277,602

TOY FIGURE HAVING MOVABLE JOINTS

Filed June 15, 1964 2 Sheets-Sheet 1

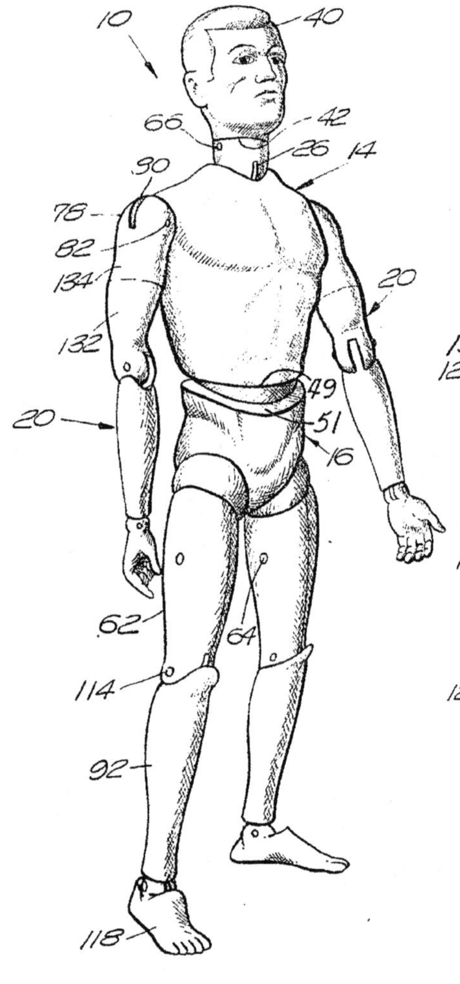

FIG. 1

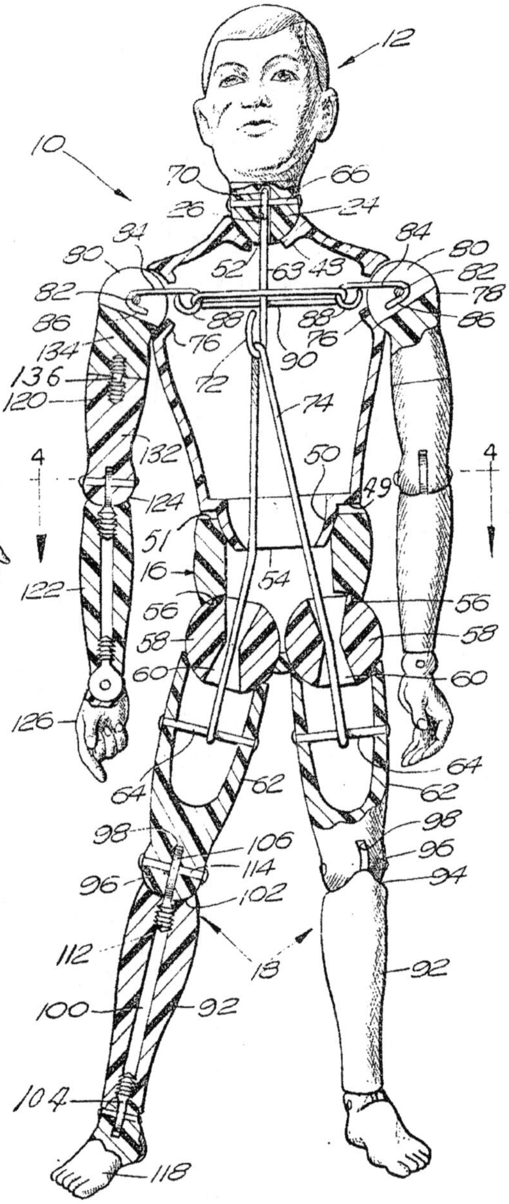

FIG. 2

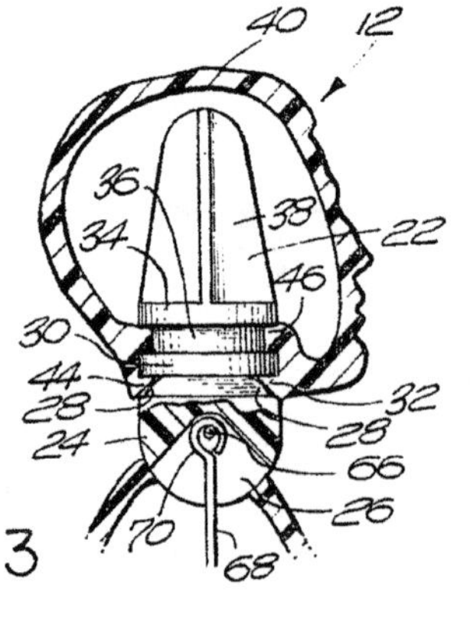

FIG. 3

INVENTORS
SAMUEL F. SPEERS
HUBERT P. O'CONNOR
BY
Salter & Michaelson
ATTORNEYS

WOWidea! Boys would never play with dolls! Samuel Speers and Hubert O'Connor teamed up to show skeptics that boys could in fact, use male dolls in their playtime fantasies. As toy designers at Hasbro, the two created a first-of-its-kind movable action figure, G.I. Joe®, with clothes and accessories that matched his adventures.

Jan. 24, 1950 P. L. SPENCER **2,495,429**

METHOD OF TREATING FOODSTUFFS

Filed Oct. 8, 1945

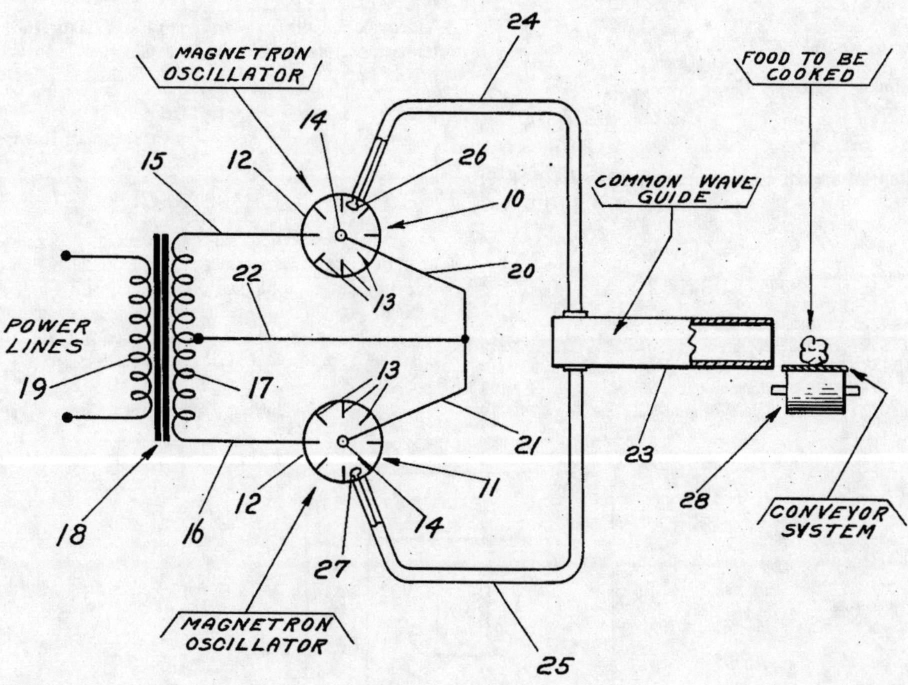

INVENTOR.
PERCY L. SPENCER,
BY Elmer J. Gorn
ATTY.

WOWidea! Turned combat into cooking, by accident! Percy Spencer, a noted World War II innovator within the field of military radar systems at Raytheon Corporation, happened to be in the radar lab when a candy bar melts in his pocket. He transformed this "Aha!" experience into the first microwave oven, known then as the Radarange®.

US005493618A

United States Patent [19]

Stevens et al.

[11] **Patent Number: 5,493,618**

[45] **Date of Patent: Feb. 20, 1996**

[54] **METHOD AND APPARATUS FOR ACTIVATING SWITCHES IN RESPONSE TO DIFFERENT ACOUSTIC SIGNALS**

[75] Inventors: **Carlile R. Stevens**, Horseshoe Bay, Tex.; **Dale E. Reamer**, Lafayette, Calif.

[73] Assignee: **Joseph Enterprises**, San Francisco, Calif.

[21] Appl. No.: **58,727**

[22] Filed: **May 7, 1993**

[51] **Int. Cl.**^{6} **H04B 1/00**

[52] **U.S. Cl.** **381/110**; 381/56

[58] **Field of Search** 381/110, 56, 7; 367/197–199

[56] **References Cited**

U.S. PATENT DOCUMENTS

D. 299,127 12/1988 Boguss .
4,207,959 6/1980 Youdin et al. 381/110
4,513,189 4/1985 Ueda et al. 381/110
4,641,292 2/1987 Tunnell et al. 367/198
4,856,072 8/1989 Schneider et al. 381/110
5,199,080 3/1993 Kimura et al. 381/110

FOREIGN PATENT DOCUMENTS

1250654 2/1989 Canada .
3608497 9/1987 Germany 381/110

OTHER PUBLICATIONS

Product Advertisement for The Clapper™, Joseph Enterprises, Inc.

Videotape of thirty (30) second and sixty (60) second television commercials for The Clapper™, Joseph Enterprises, Inc.

Primary Examiner—Stephen Brinich
Attorney, Agent, or Firm—Townsend and Townsend and Crew

[57] **ABSTRACT**

An acoustic switch device that independently operates two or more electrical appliances. The acoustic switch operates a first electrical appliance upon receipt of a first series of acoustic signals and operates a second electrical appliance upon receipt of a second series of acoustic signals that is different from the first series of acoustic signals.

9 Claims, 5 Drawing Sheets

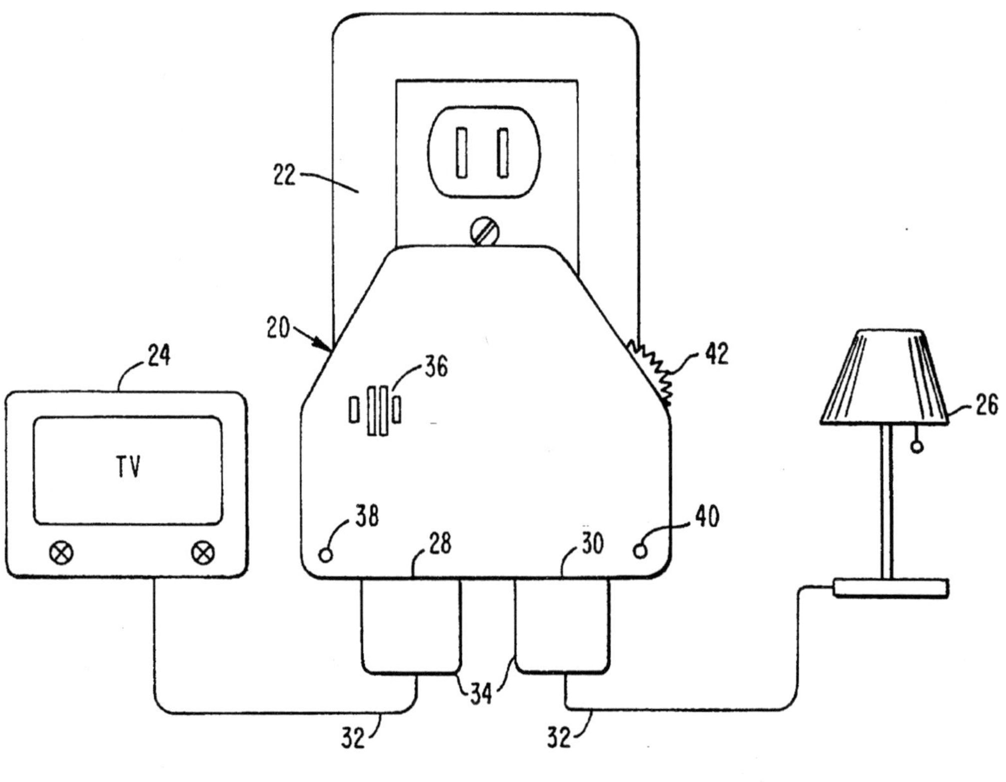

WOWidea! Clap on, clap off! When Carlile Stevens and Dale Reamer invented their Clapper®, little did American television viewers realize they would soon be reciting the mesmerizing mantra at dinner parties and around the water cooler at work. It surely joins, "I've fallen and can't get up!", in the TV-commercial hall of fame.

DESIGN.

No. 37,508. PATENTED AUG. 8, 1905.

G. STICKLEY.

CHAIR FRAME.

APPLICATION FILED JUNE 26, 1905.

WITNESSES:

B. E. Robinson.

M. M. Nott.

INVENTOR:

Gustav Stickley

BY:

Howard P. Denison

ATTORNEY.

ANDREW. B. GRAHAM CO., PHOTO-LITHOGRAPHERS, WASHINGTON, D. C.

WOWidea! Simplicity is timeless! Gustav Stickley, is the best known of several furniture-making brothers who introduced the European Arts and Crafts movement to the U.S. They also brought along the concept of utopian work communities. During the difficult economy of World War I, the original company was reorganized and survived.

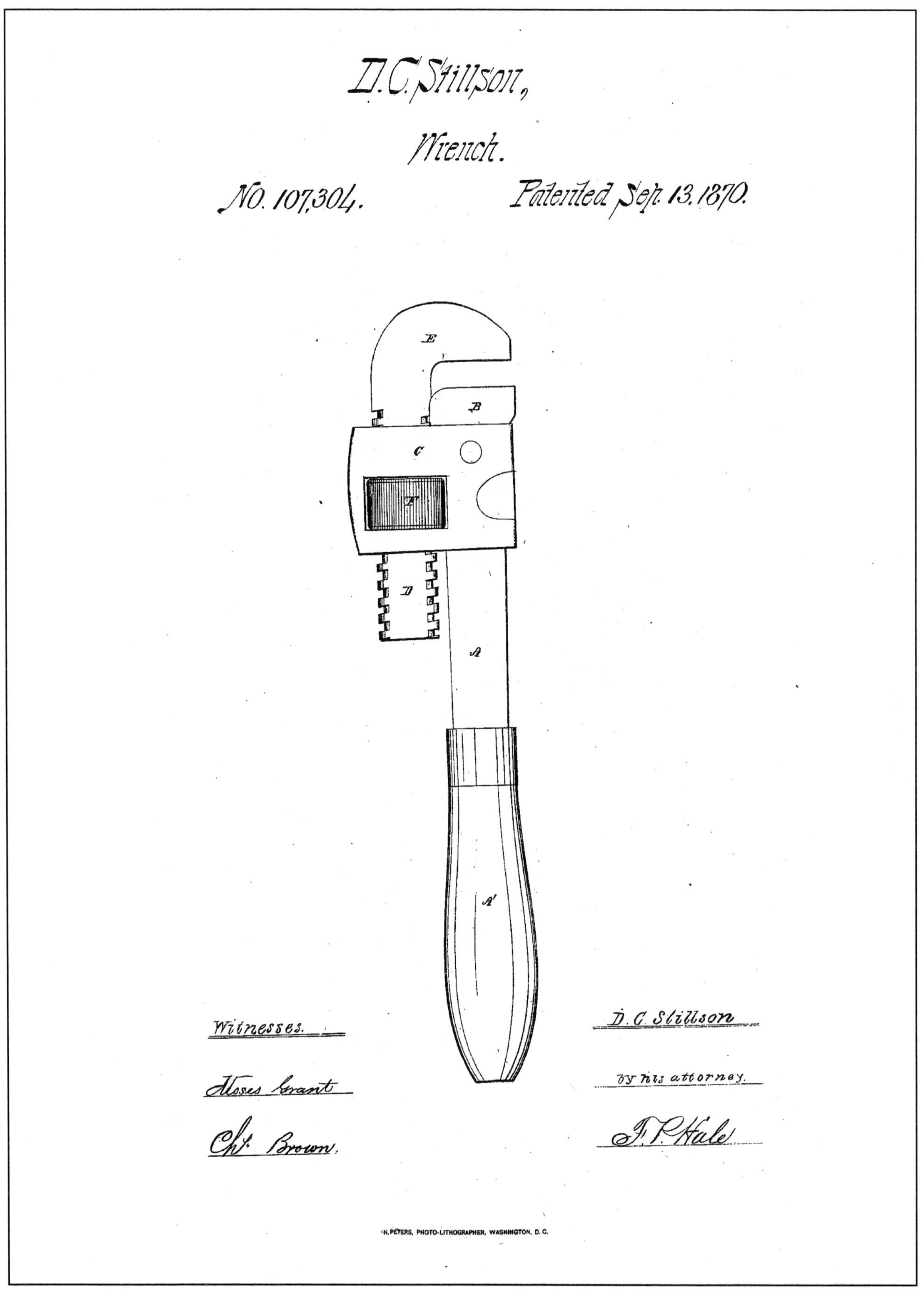

WOWidea! Leverage your mind! Daniel Stillson, a steamboat fireman, suggested to Walworth Manufacturing a lever-based design for a wrench that could be used for screwing pipes together. Previously, serrated blacksmith tongs had been used for that purpose. His thoughts earned him about $80,000 in royalties during his lifetime.

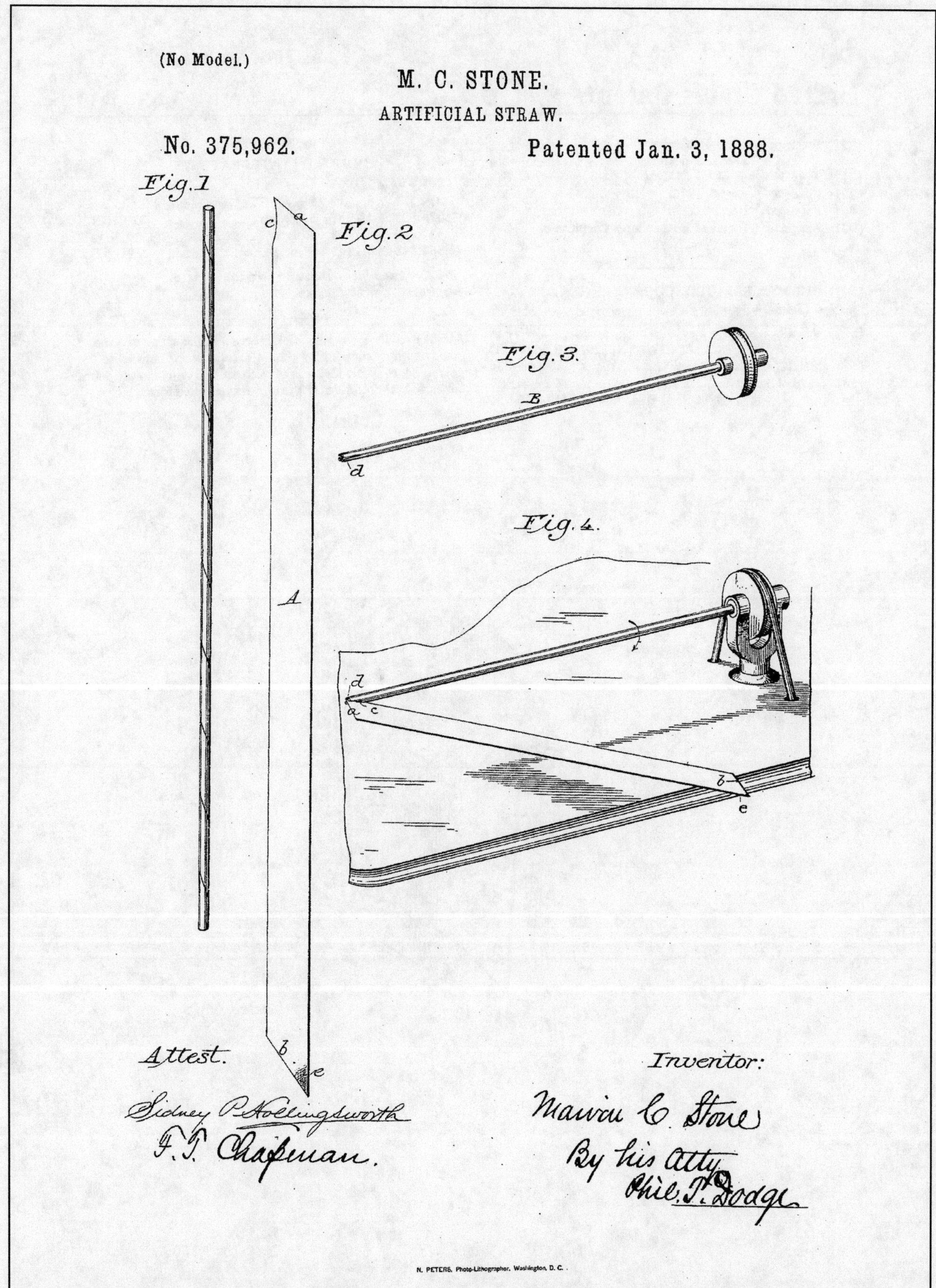

WOWidea! Mint-julep drinks might have provided the inspiration! Marvin Stone's spiral-winding process was used to manufacture the first paper drinking straws. His company, Stone Industrial, first made cigarette holders but soon he got the idea of making drinking straws from paper as a substitute for the rye grass straws used at the time.

United States Patent

[11] 3,630,430

[72] Inventor **Glenn E. Struble**
Fairfield, Ohio
[21] Appl. No. **61,621**
[22] Filed **Aug. 6, 1970**
[45] Patented **Dec. 28, 1971**
[73] Assignee **Diamond International Corporation**
New York, N.Y.

[54] **QUICKLY ERECTED SCOOP-TYPE CARTON**
6 Claims, 4 Drawing Figs.

[52] **U.S. Cl.** .... **229/16 B,** 229/1.5 B, 229/41 B, 229/41 D
[51] **Int. Cl.** **B65d 5/36**
[50] **Field of Search** .... 229/21, 1.5 B, 16 R, 41 R, 41 B, 41 C, 41 D; 294/55

[56] **References Cited**

UNITED STATES PATENTS

904,050	11/1908	Crawford	229/21
1,034,522	8/1912	Shaw	229/1.5 B UX
2,078,038	4/1937	Stephens	229/16 R X
2,337,199	12/1943	Holy	229/16 R
2,385,898	10/1945	Waters	229/1.5 B X

Primary Examiner—Donald F. Norton
Attorney—Karl W. Flocks

ABSTRACT: A box for quick erection in the shape of a scoop made with application of glue in parallel strips parallel to the blank edges and with an arcuate bottom having curved score lines and tapered sides to the box in assembled form.

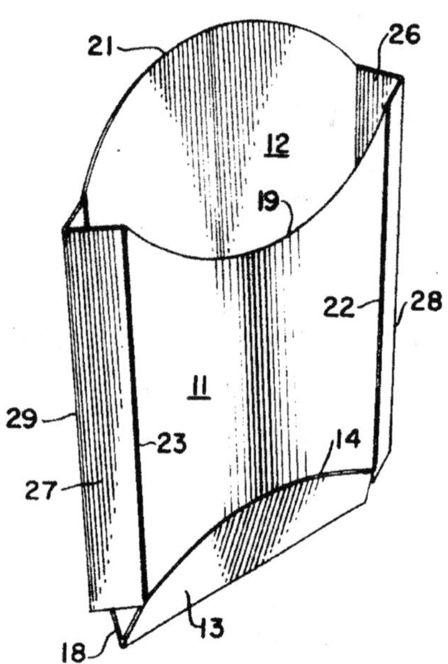

WOWidea! Well, someone's got to invent it! Glenn Struble is not a name normally associated with greatness, but neither are most inventors. The vast majority of everyday inventions are created and patented by common people looking for uncommon solutions. What did he dream up? The fast food container for McDonald's® French fries.

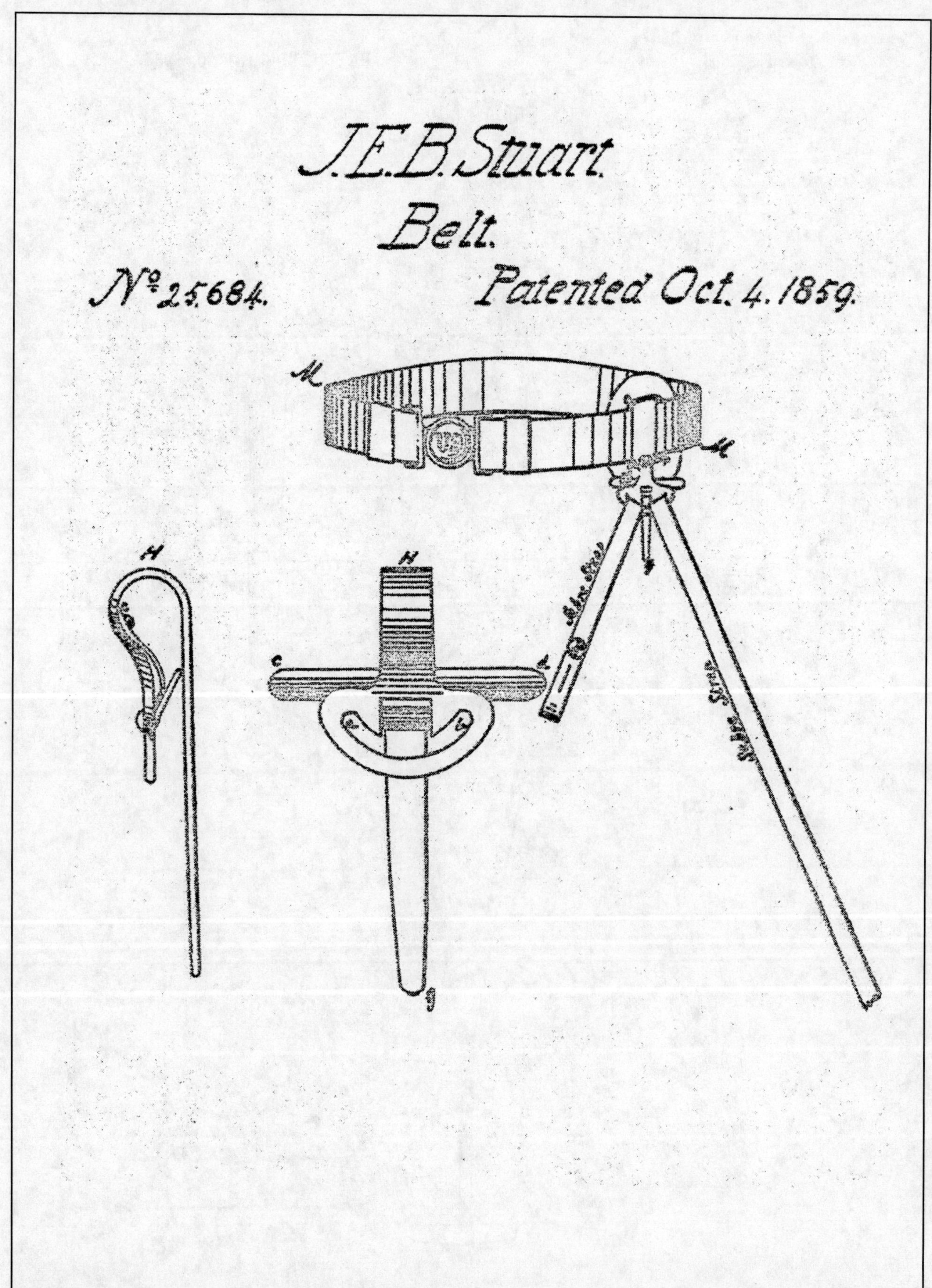

WOWidea! Start with what you know! James Ewell Brown Stuart (JEB Stuart) invented this sword holder for the cavalry and tried selling it to the U.S. government. Before switching sides in the Civil War, he graduated 13th in his class at West Point, which is where he met his friend and future Confederate army commander, Robert E. Lee.

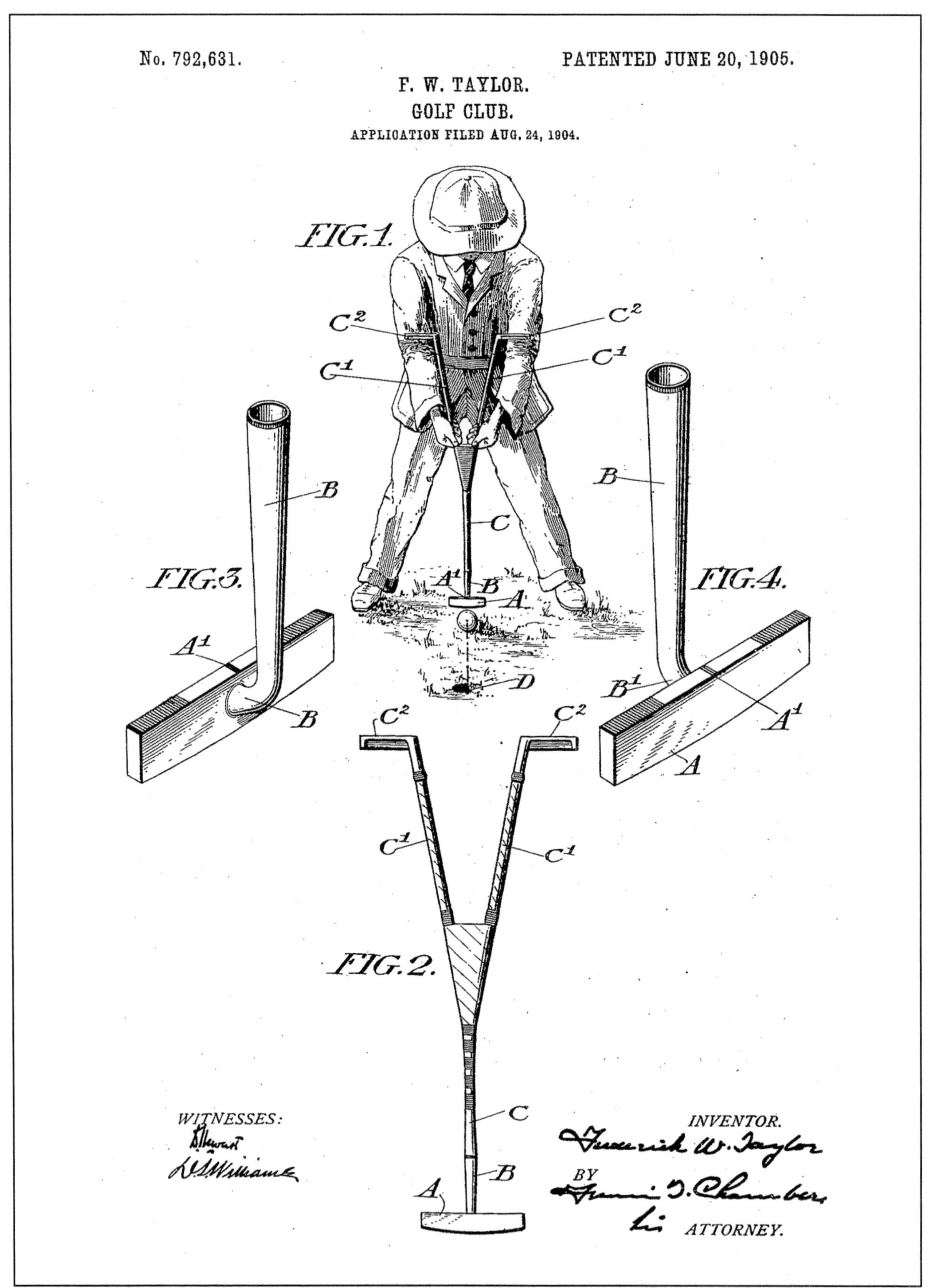

WOWidea! Time and motion studies go golfing! At the beginning of the 20th Century, Frederick Winslow Taylor was considered the guru of "scientific management" and Henry Ford used Taylor's ideas almost exclusively. So when he began to apply his theories to golf he invented this putter, so accurate, it was banned from American golf courses.

(No Model.) 4 Sheets—Sheet 2.

N. TESLA.

ELECTRO MAGNETIC MOTOR.

No. 381,968. Patented May 1, 1888.

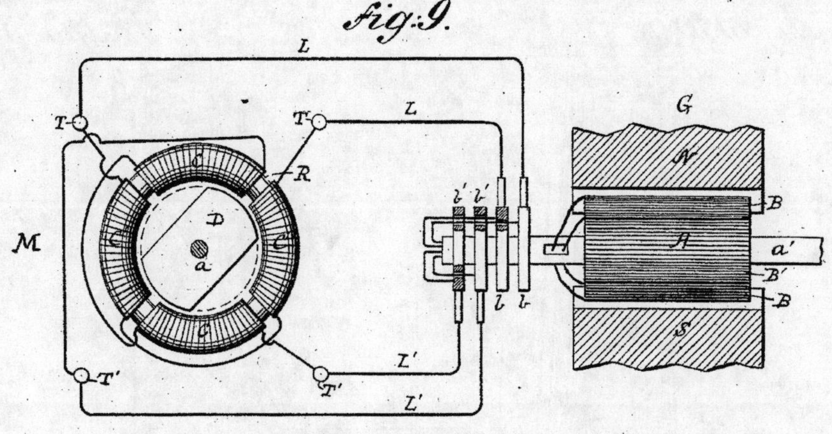

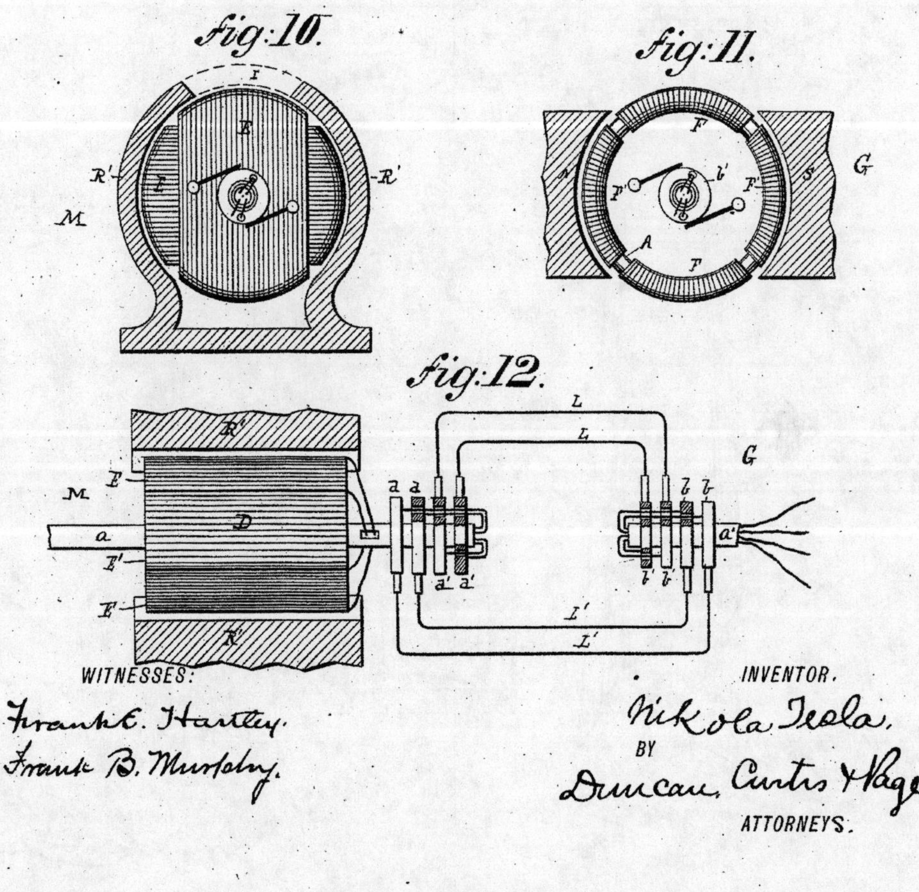

N. PETERS, Photo-Lithographer, Washington, D. C.

WOWidea! Rotating magnetic fields create power! The first commercially viable AC motor sprung from the mind of Croatian-born Tesla and led to his infamous rift with his employer, Thomas Edison. Recognized as a genius by Edison's nemesis, George Westinghouse, Tesla's designs changed the world, yet he ultimately died in obscurity.

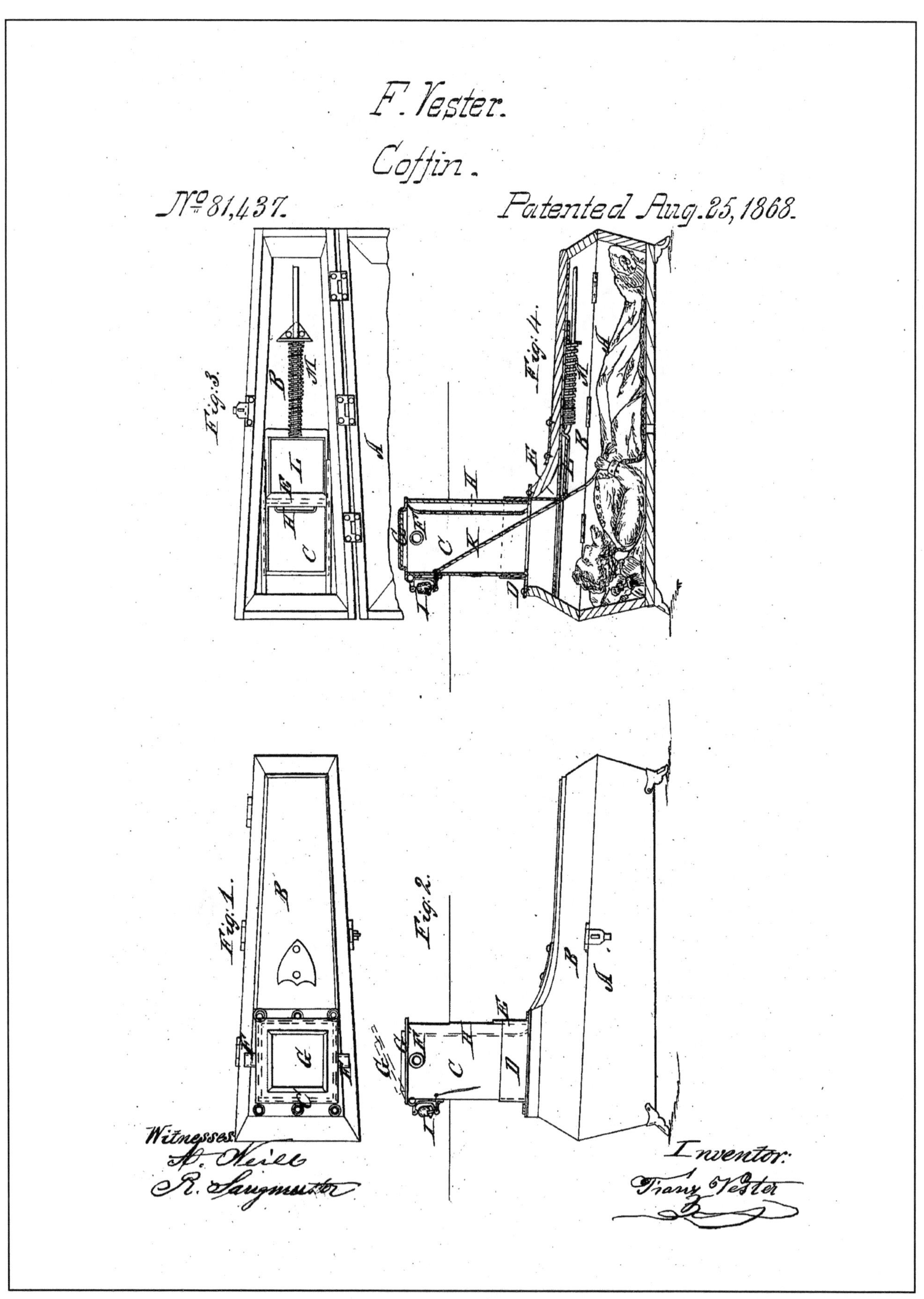

WOWidea! Climb out or ring the bell if you find yourself buried alive! Only recently have we had the necessary medical understanding, devices and procedures to determine if a person was truly dead. The Irish didn't call it "waking" without reason. Franz Vestor was one of many to invent a method for escape in case of mistake.

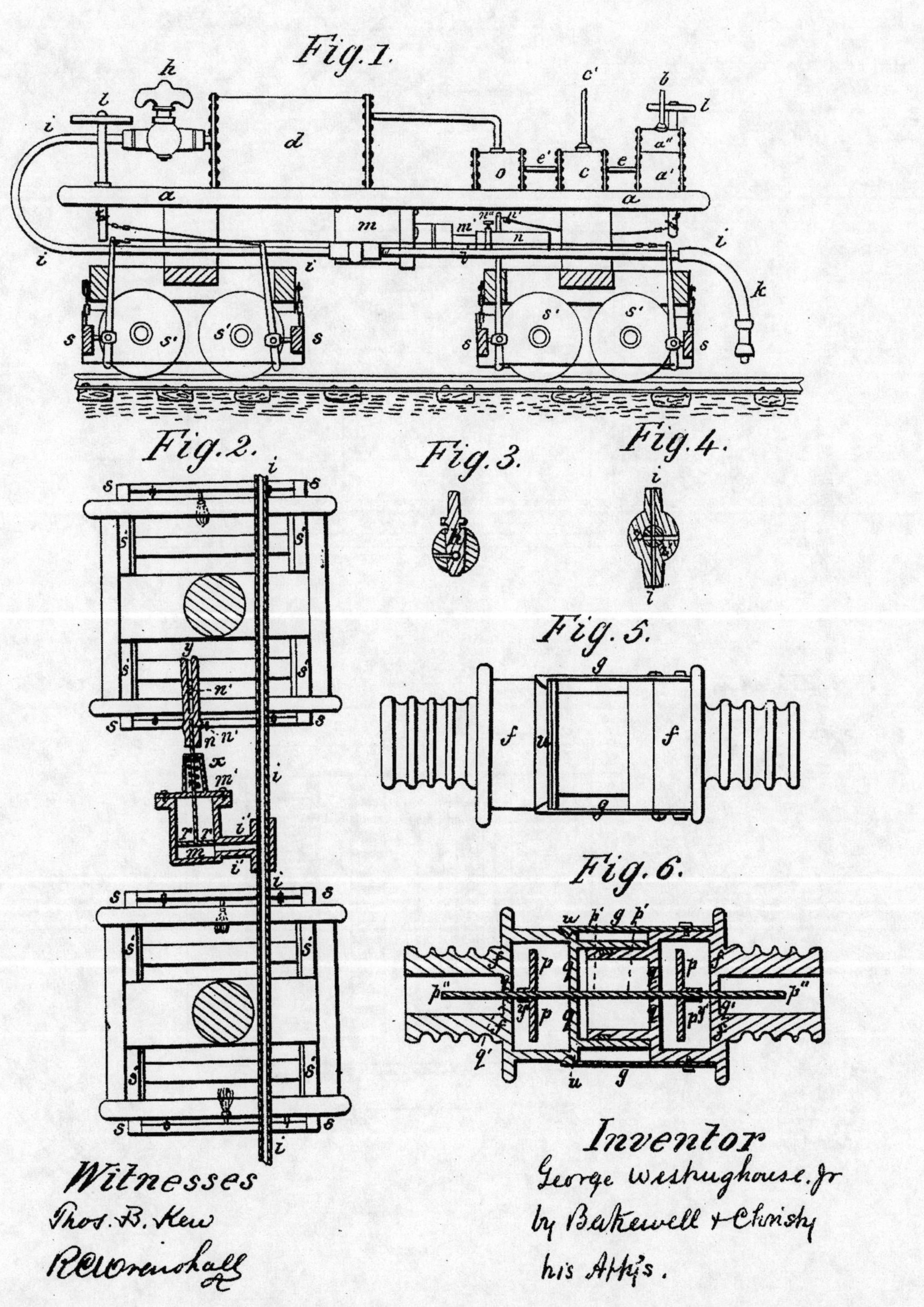

WOWidea! Centrally controlled compressed air safely brought speeding trains to a safe stop like never before! A college dropout, Westinghouse was fascinated with train wrecks, patenting a train re-railer at 21, then going on to create an enormous empire built on innovations in railroading, natural gas and AC electrical power systems.

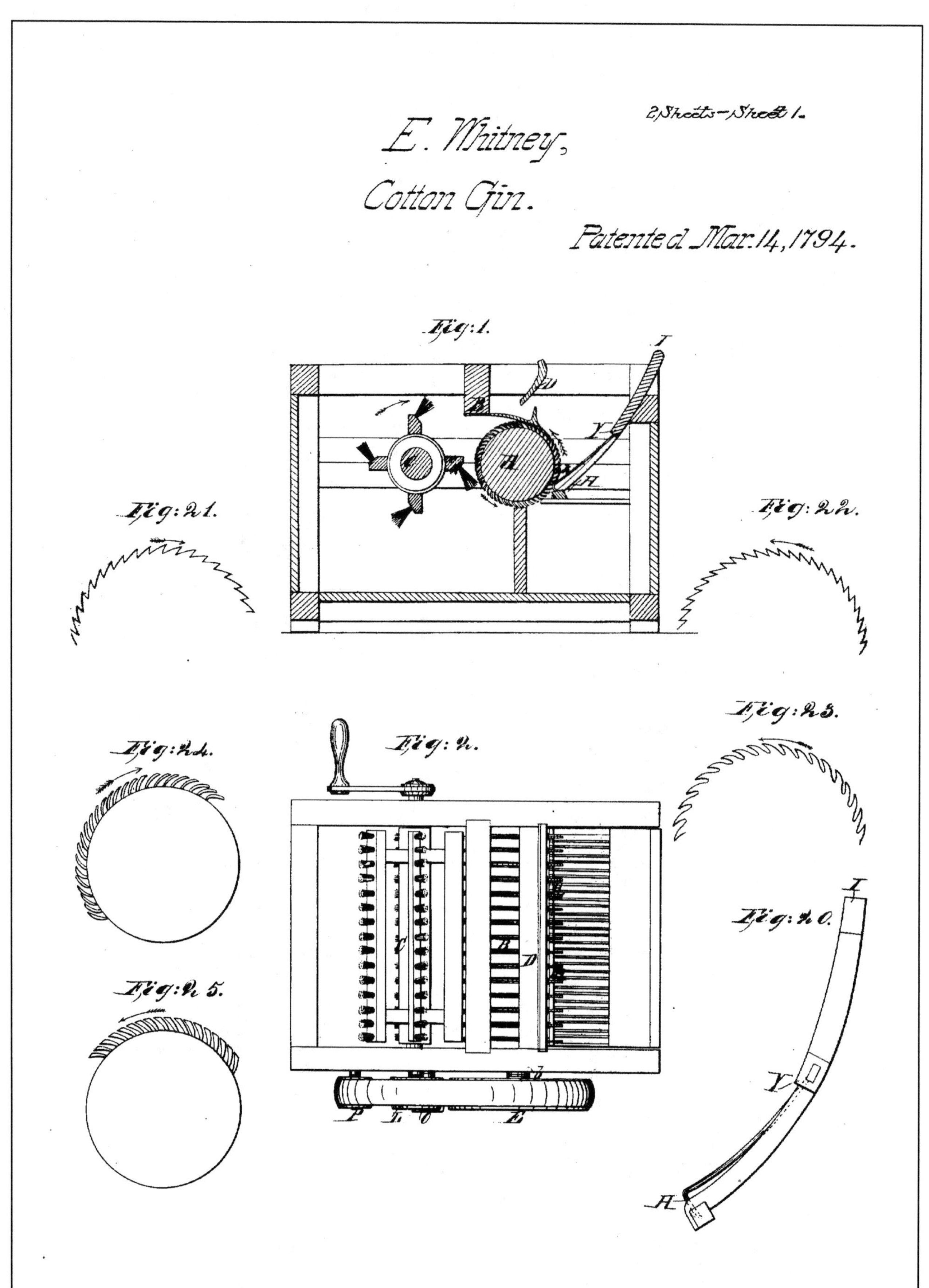

WOWidea! Revolutionary in every way! With rotating rollers, this "Gin", short for engine, separated cotton from its husk. Invented by Eli Whitney after meeting the widow of an American Revolutionary War hero (she owned a large southern cotton plantation), it was unfortunately copied by others, so Whitney later found his fortune in firearms.

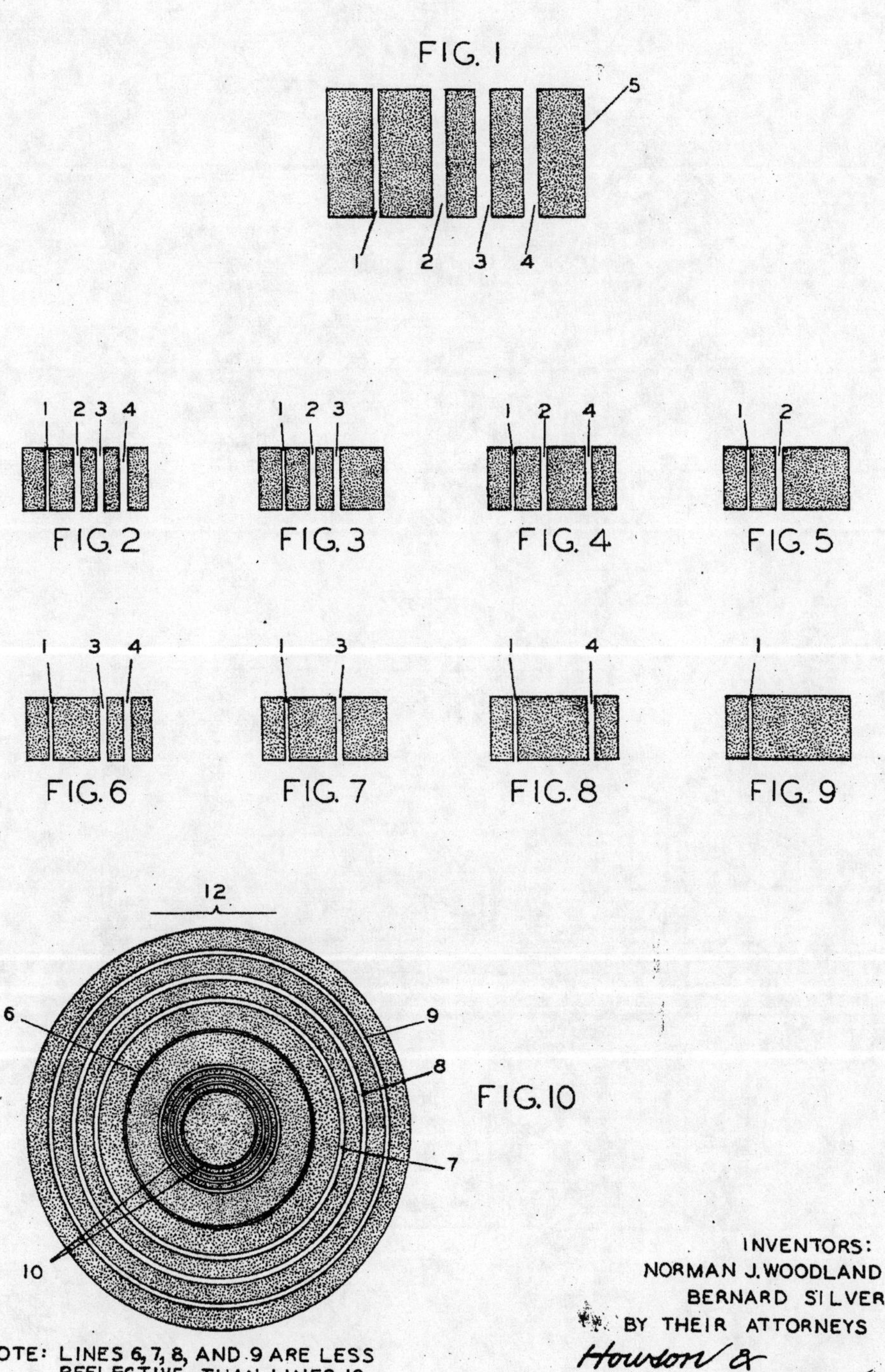

WOWidea! Pattern born out of "extending" Morse code! Norman Woodland and Bernard Silver were graduate students when they overheard that a local food store chain was interested in automatically capturing product information at check-out. After trial and error, bullseye became barcode. First product scanned? Chewing gum.

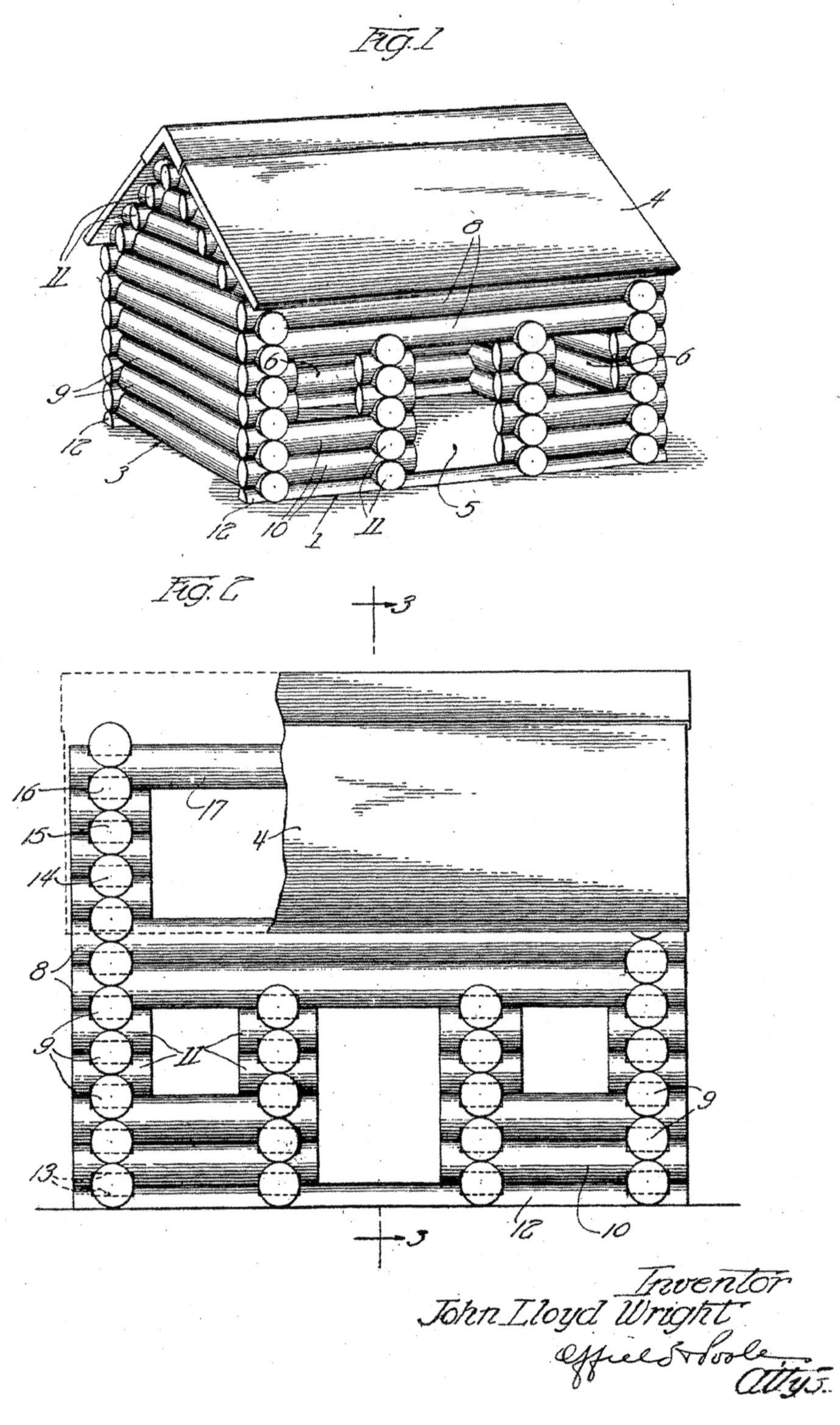

WOWidea! Earthquakes toyed with his mind! John Loyd Wright was inspired when he joined his architect father, Frank Loyd Wright, on a trip to Japan to oversee the building of his quake-proof design used in Tokyo's Imperial Hotel. Lincoln Logs® were so named to evoke a connection to the log cabin of Abraham Lincoln's childhood.

No. 821,393. PATENTED MAY 22, 1906.

O. & W. WRIGHT.

FLYING MACHINE.

APPLICATION FILED MAR. 23, 1903.

3 SHEETS—SHEET 1.

FIG. 1.

WITNESSES:

William F. Bauer.

Irvine Miller.

INVENTORS.

Orville Wright

Wilbur Wright

BY H. A. Toulmin,

ATTORNEY.

WOWidea! Warped wings helped lift the load! The Wright brothers were high school dropouts, bicycle shop owners and most importantly, mechanical contraption enthusiasts. They eventually convinced the skeptical U.S. Patent Office that they invented the first "controlled, man-carrying, heavier-than-air, motor-driven, flying machine."

United States Patent [19]

Wyeth et al.

[11] **3,935,358**

[45] **Jan. 27, 1976**

[54] **PROCESS FOR PREPARING A HOLLOW, RIB-REINFORCED LAMINATED STRUCTURE**

[75] Inventors: **Nathaniel Convers Wyeth,** Mendenhall, Pa.; **Frank William Arnoth,** Wilmington, Del.

[73] Assignee: **E. I. Du Pont de Nemours & Co.,** Wilmington, Del.

[22] Filed: **Feb. 15, 1974**

[21] Appl. No.: **443,199**

Related U.S. Application Data

[63] Continuation-in-part of Ser. No. 346,164, March 29, 1973, abandoned.

[52] **U.S. Cl.** .... **428/166;** 156/245; 156/285; 264/89; 264/92; 264/94; 264/248; 264/296; 425/387 B; 425/388; 425/503; 425/504; 428/167

[51] **Int. Cl.²**. **B29C 17/04;** B32B 3/00; B32B 31/20

[58] **Field of Search** .... 264/89, 90, 92, 93, 94, 264/96, 98, 99, 294, 296, 248; 156/145, 221, 156, 285, 292, 245; 29/157.3 V, 421 R; 72/60–62; 425/387, 388, 387 B, 503, 504; 161/122, 123; 428/166, 167

[56] **References Cited**

UNITED STATES PATENTS

3,106,014	10/1963	Brick et al.	29/157.3 V UX
3,141,913	7/1964	Edwards	264/296 X
3,271,846	9/1966	Buechele et al.	29/421 X
3,281,301	10/1966	Bolesky	264/94 X

Primary Examiner—Jan H. Silbaugh

[57] **ABSTRACT**

The invention provides a process for preparing a hollow, rib-reinforced, laminated article by:

a. placing two sheets between opposing mold platens, the sheets being aligned such that the sheet surfaces oppose each other, at least one of the sheets being a thermoplastic material heated to its thermoforming temperature and at least one of the sheets being provided with grooves or integral projections which form fluid passageways, and at least one of the mold platens being provided with a mold caivty to form a shaped article having ribs;

b. closing the mold platens to contact the sheets; and

c. introducing a fluid into the fluid passageways to distend the thermoplastic sheet into its mold cavity forming a shaped article having ribs, while the sheets maintain contact in the nondistended areas.

34 Claims, 16 Drawing Figures

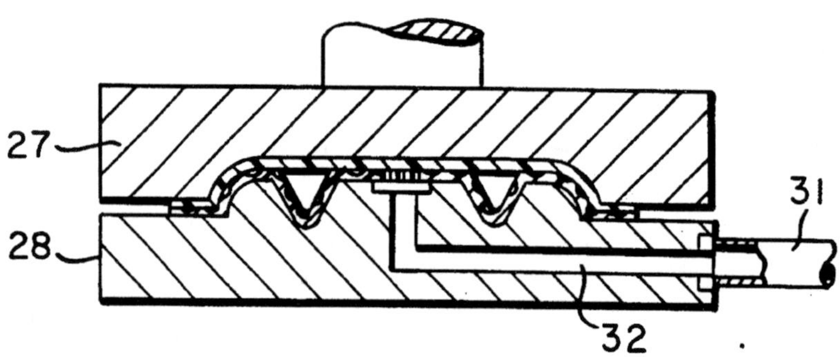

WOWidea! Stretching nylon molecules in two dimensions, made bottles resist the carbonated pop! Born into an artistic dynasty, Nathaniel Wyeth, along with his collaborator, sought mechanical utopia by creating the ubiquitous plastic soda bottle that has transformed the packaging industry and ultimately, the recycling business too.

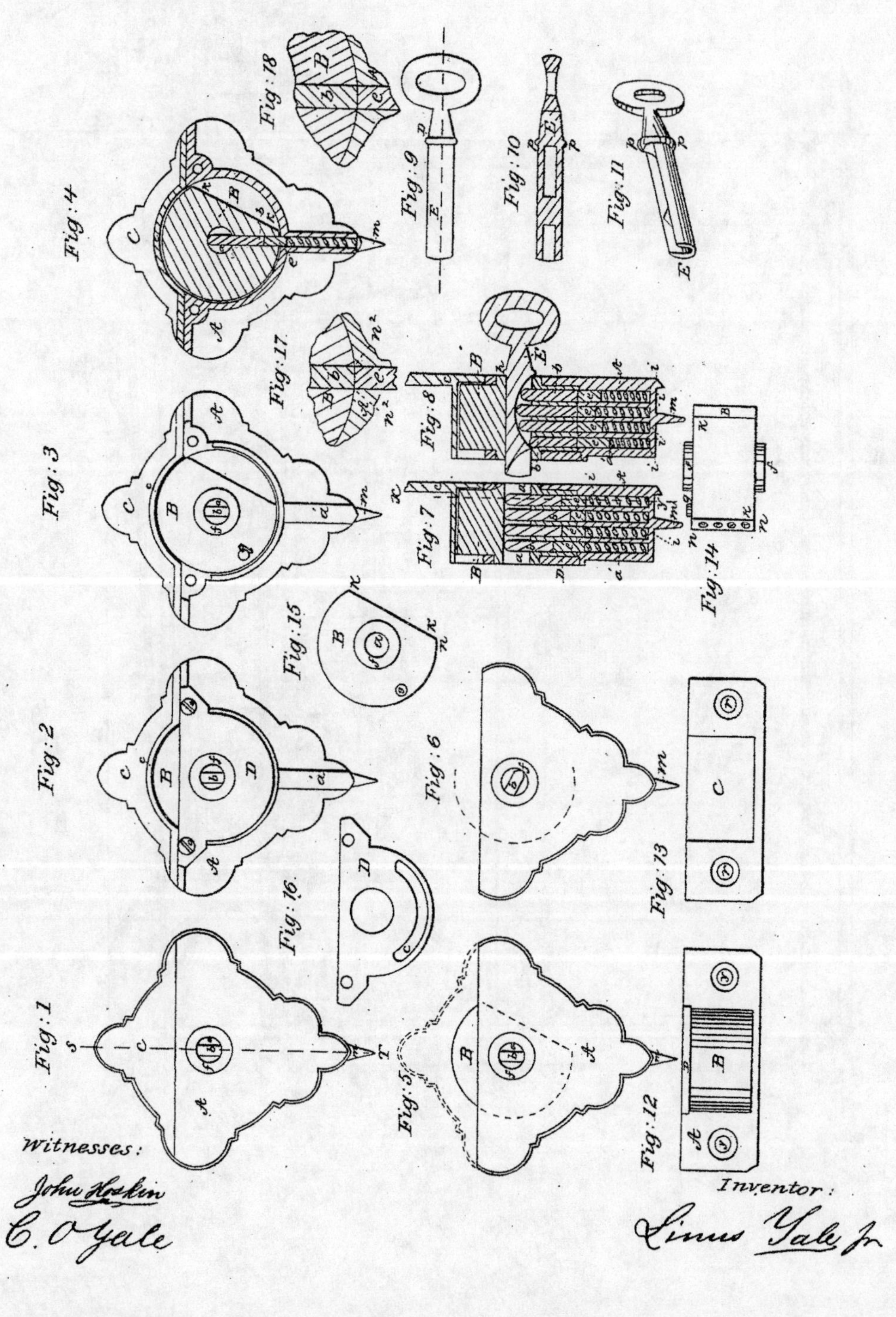

WOWidea! Thin flat key combines with pin-tumbler to lock out the competition! Son of an inventive father, Linus Yale Jr., excelled at mechanical contrivances yet lacked good business sense. For that, he partnered with John Towne, creating a lock legacy that would long outlast him; Linus died 2 months into the fledgling partnership.

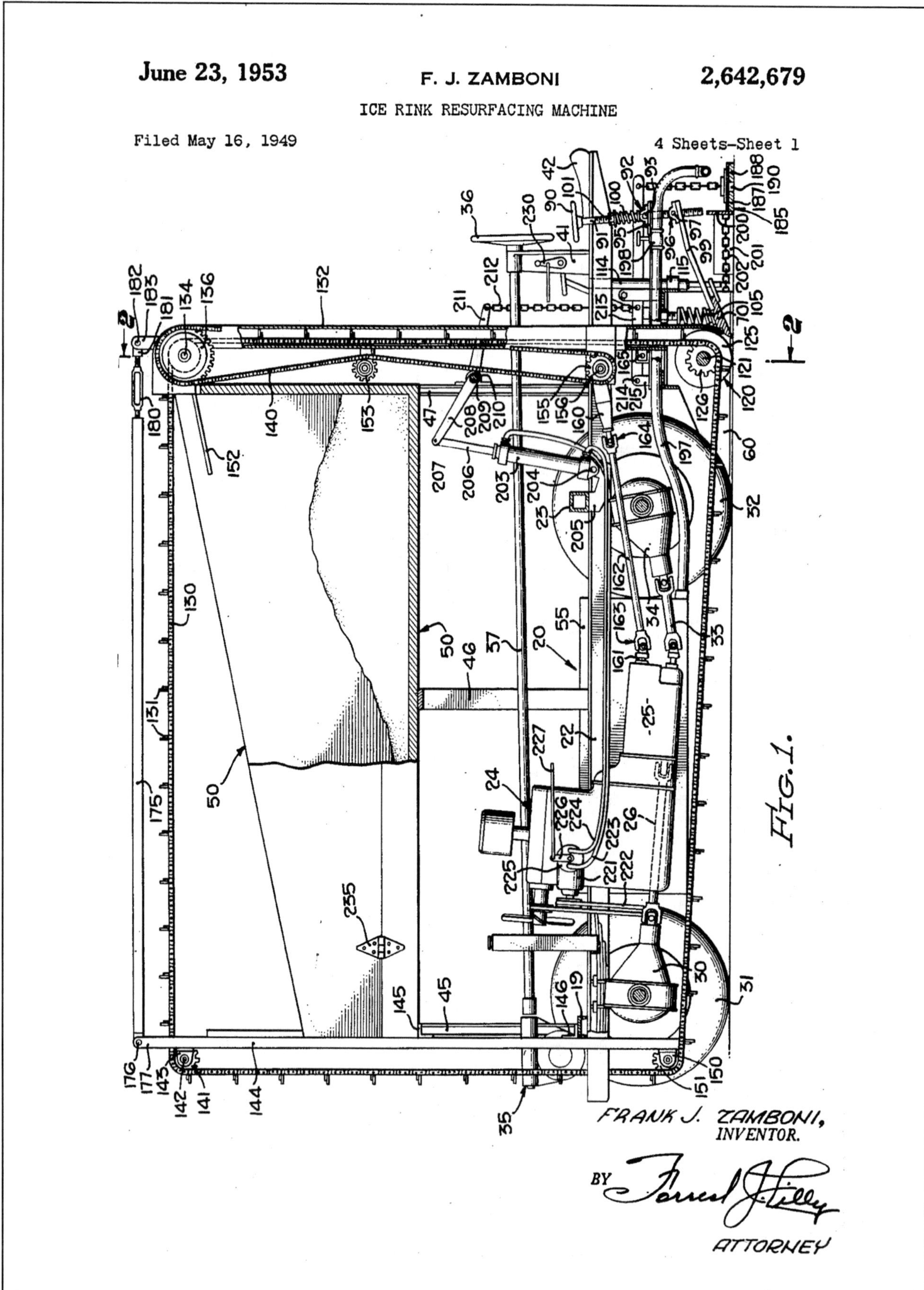

WOWidea! He wanted it like glass! When Frank Zamboni moved to Southern California he began by selling blocks of ice, but business soon melted when freezer technology heated up. Looking to use his new expertise, he opened a skating rink and found keeping the icy surface smooth was a tough job- the Zamboni® machine was born.

No. 621,195. Patented Mar. 14, 1899.

FERDINAND GRAF ZEPPELIN.

NAVIGABLE BALLOON.

(Application filed Dec. 29, 1897.)

(No Model.) 4 Sheets—Sheet 1.

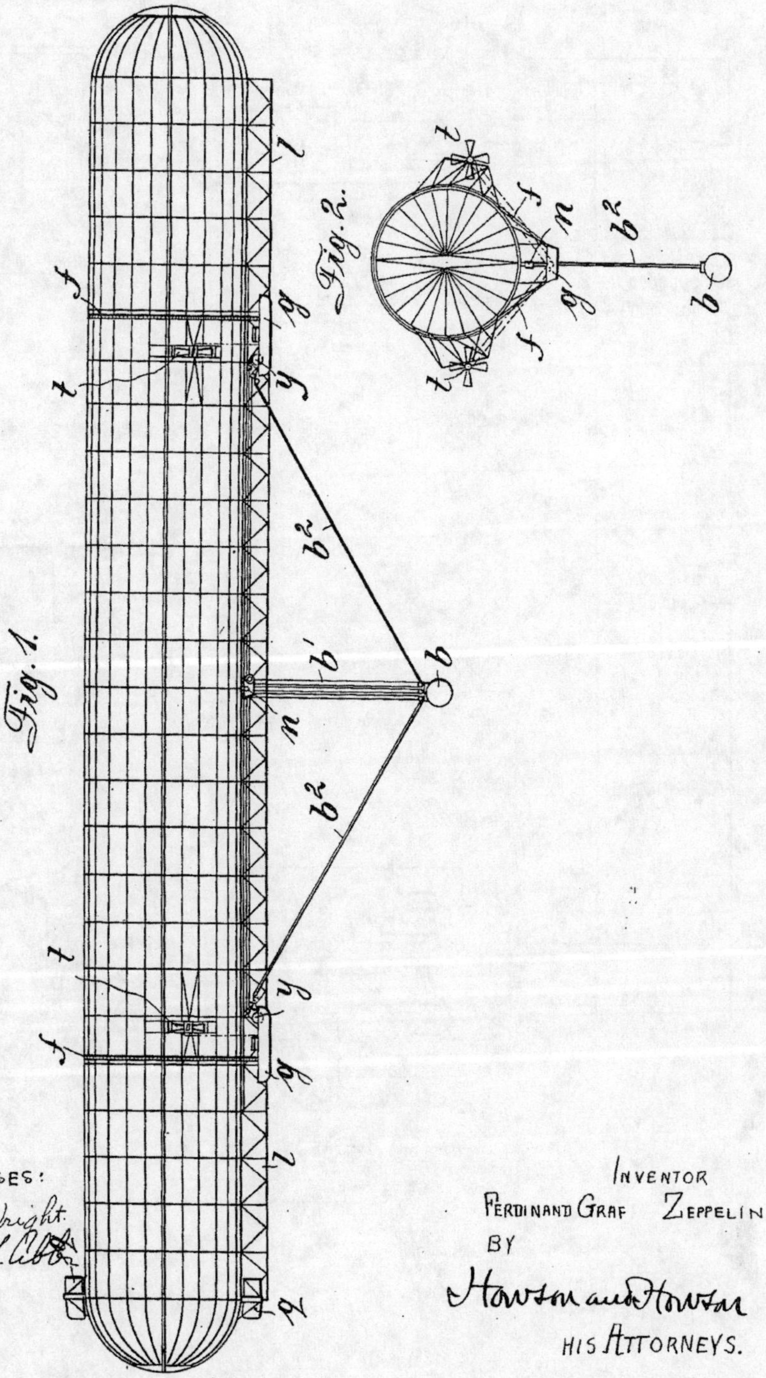

WITNESSES:
F. A. Wright.
H. L. Cibb

INVENTOR
FERDINAND GRAF ZEPPELIN
BY
Howson and Howson
HIS ATTORNEYS.

WOWidea! Balloons lifted his thoughts! Ferdinand Zeppelin, born in Germany, attended military academy and soon became an officer. In 1863 he traveled to the United States to act as an observer for Union forces and by his late 50's he had developed an interest in airships. In 1910, his design carried the first commercial air passengers.

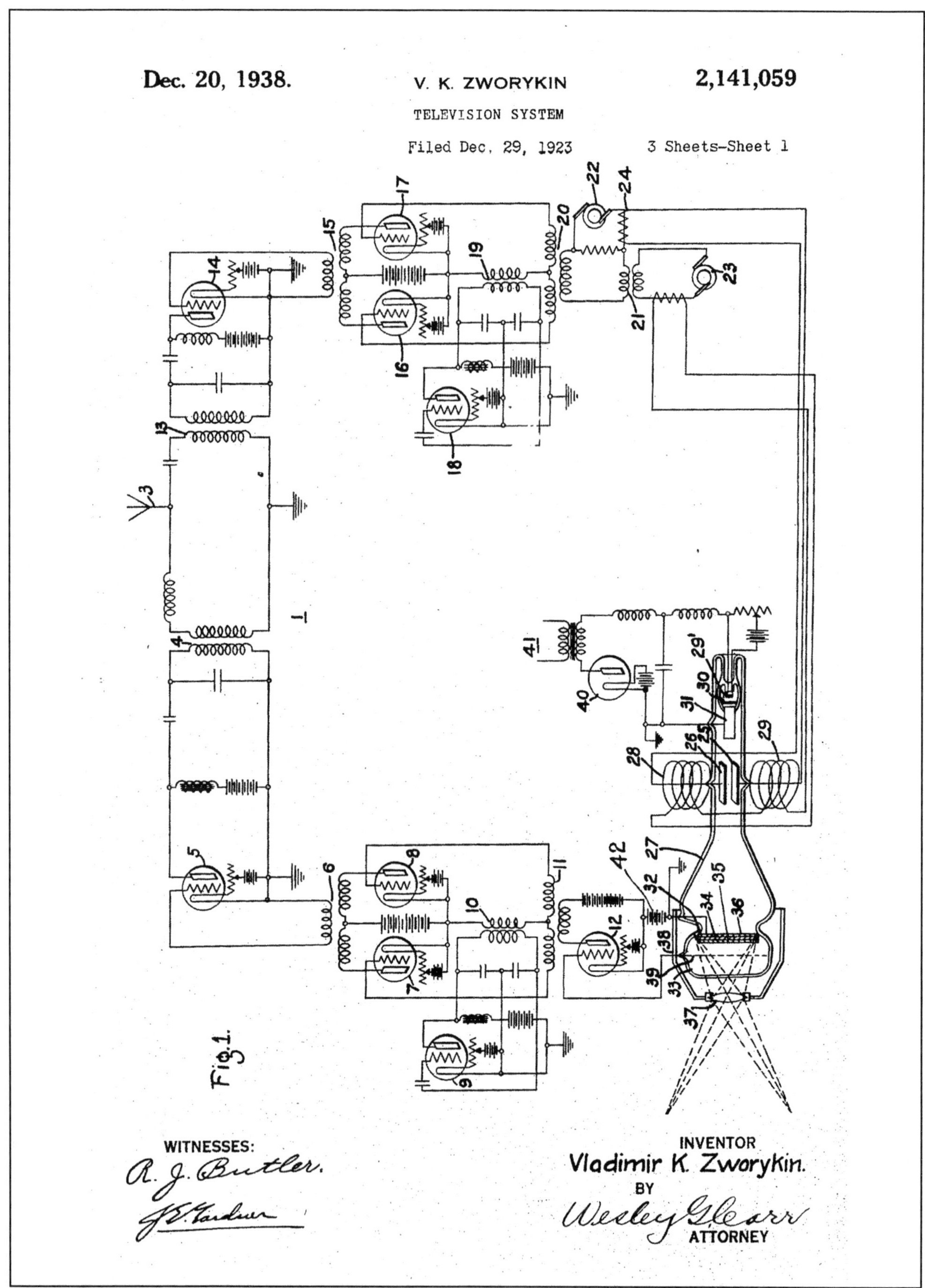

WOWidea! Patience is profitable! Vladimir Zworykin, a naturalized U.S. citizen once worked as a bookkeeper in the Russian Embassy. His great contribution to the modern television industry was the image-gathering and image-displaying cathode-ray tubes. The patent application was fought by a competing inventor, Farnsworth, for 15 years.

Further Research…

There were untold resources used in developing this extensive collection. In addition to the selected sources listed on the next page, many museums, along with the Internet, have vast stores of knowledge related to patents, inventions and the act of creativity. In particular, you are encouraged to start your explorations by visiting the places that guided us in our initial quest for WOWideas!

•The United States Patent and Trademark Office, Crystal City, Virginia. www.uspto.gov

•The National Inventors Hall of Fame, Akron, Ohio. www.invent.org

•The Henry Ford Museum, Dearborn, Michigan. www.thehenryford.org

•The Tech Museum of Innovation, San Jose, California. www.thetech.org

•The Smithsonian National Museum of American History, Washington, DC. www.si.edu

•The Mercer Museum, Doylestown, Pennsylvania. www.mercermuseum.org

•The Museum of Science and Industry, Chicago, Illinois. www.msichicago.org

•The Franklin Institute, Philadelphia, Pennsylvania. www.fi.edu

•The Edison National Historic Site, West Orange, New Jersey. http://www.nps.gov/edis/

And finally, for a complete listing of over 400 fun, hands-on, experiential Science Centers in over 40 countries, please visit www.tryscience.org, sponsored by the Association of Science-Technology Centers.

Selected Sources

Bourne, Russell (1996). Invention in America. Golden, CO: Fulcrum Publishing

Brown, David (2002) Inventing Modern America: from microwave to the mouse. Boston, MA: Massachusetts Institute of Technology

Brown, Travis (1994). Historical First Patents: the first United States patent for many everyday things. Metuchen, NJ: The Scarecrow Press, Inc.

Editors, American Heritage of Invention & Technology (1995). American Inventions: a chronicle of achievements that changed the world. New York, NY: Barnes & Noble Books

Editors, Inventors and Discoverers: changing our world (1988). Washington, DC: The National Geographic Society

Editors, The Smithsonian Book of Invention. Washington, DC: Smithonsian Expedition Books

Flatow, Ira (1992). They All Laughed.... from light bulbs to lasers: the fascinating stories behind the great inventions that have changed our lives. New York, NY: HarperCollins Publishers

Ierley, Merritt (2002). Wondrous Contrivances: technology at the threshold. New York, NY: Clarkson Potter- Publishers

Karwatka, Dennis (1996). Technology's Past Volume 1. Ann Arbor, MI: Prakken Publications, Inc.

Karwatka, Dennis (1999). Technology's Past Volume 2. Ann Arbor, MI: Prakken Publications, Inc.

Lasson, Kenneth (1986). Mousetraps and Muffling Caps. New York, NY: Arbor House Publishing Company

Lienhard, John H. (2000). The Engines of our Ingenuity: an engineer looks at technology and culture. New York , NY: Oxford Uniersity Press

National History Day Home Page. (1999). History of the patent office. <http://www.nationalhistoryday.org/03_educators/teach99/lesson2/lesson2_text.htm>

Perdue University Libraries THOR Home Page. (1999). Types of intellectual property. <http://gemini.lib.purdue.edu/instruction/gs175/Spring99/gs175d/lesson_vi.html>

Petroski, Henry (1996). Invention by Design: how engineers get from thought to thing. Cambridge, MA: Harvard Univeristy Press

Van Dulken, Stephen (2000). Inventing the 20th Century: 100 inventions that shaped the world: from the airplane to the zipper. Bury St Edmunds, England: St Edmundsbury Press

Van Dulken, Stephen (2001). Inventing the 19th Century: 100 inventions that shaped the victorian age: from aspirin to the Zeplin. New York, NY: New York University Press

In Closing…

We hoped you've enjoyed this collection. To obtain bulk copies of *this* paperback book or to provide feedback please contact us (see below).

Customized Limited Edition Available

Do you know an Inventor? Are you an Inventor? Was there a patented invention in your family or company's history? Our customized leather-bound edition makes the perfect personal, family or corporate gift. We can insert, alphabetically by inventor's last name, one or more United States patents of *your* choosing into a masterfully crafted edition of WOWideas! Each book is unique, and can be signed and numbered.

Simply provide us with the Inventor's name, United States patent number and a brief history (30 words or less) of the invention. If you're missing some information, contact us anyway, we can research the United States Patent and Trademark Office files.

To obtain an order form, please phone or visit our website to email us:

Alexander Tourneu & Associates, LLC
Toll-Free 1-866-TOURNEU
www.tourneu.com

The rich leather, library binding, gold imprinted cover, red ribbon bookmark and acid-free paper make your customized collection of "WOWideas!" a timeless treasure.